Analyzing Farmer Stubble Burning Behavior

Govind Kumar

Copyright © [2023]

CONTENTS

LIST OF TABLES

LIST OF FIGURES

LIST OF PLATES

LIST OF ANNEXURES

Annexure No.	Particulars
1	Interview schedule

Introduction

Agriculture continues to be the mainstay of the Indian economy because of its greater share in employment and livelihood. Although, its relative contribution to the nation's Gross Domestic Product (GDP) has declined over the years, but still, a significant increase in area and production of crops like paddy and wheat has taken place after 1960s in North Western states of India, especially in Uttar Pradesh, Haryana and Punjab. These states represent an extremely productive paddy-wheat region in the Indo-Gangetic Plains and are called as the "food bowl of India". The Indo-Gangetic Plains in India which hold nearly 40 per cent of the total population cover about 20 per cent of the geographical area and contributes around 42 per cent to the total food grain production. In this region, around 12 million hectares is accounted for rice and wheat crop rotation. Since green revolution, the share of wheat production in the country has increased around 70-74 per cent and the share of rice to the total food grain production is around 40-45 per cent. The rice-wheat production systems in Indo-Gangetic Plains, therefore assumes paramount importance in contributing to the national pool as well as providing employment and livelihood to the millions of rural poor.

Agriculture in the states of Indo-Gangetic Plains is typically characterized by two growing seasons: a predominantly winter (*rabi*) wheat crop, harvested in April-May, and a predominantly summer (*kharif*) rice crop, harvested in October-November. Before mid-1980s, the farmers of this region used to harvest and plough fields manually. After crop harvesting, the usual practice was to leave stubbles for four to six weeks to enhance nutrients for the next crop. In the recent years, farmers have adopted mechanized harvesting due to several reasons. Studies have found that in states like Punjab, Haryana and Uttar Pradesh, more than 75 per cent of paddy is harvested using combine harvesters. Increasing utilization of mechanized harvesters over the last 30 years has decreased costs and improved efficiency for farmers. However, this harvesting method leaves more stubble on the fields than traditional methods using a sickle, and many farmers burn this residue to ready fields for the next growing season. This method of removing dry stubble by burning it to prepare the field for subsequent planting of next crop is called as stubble burning.

The remains of the field crops after harvest are of enormous use as it is a natural resource that adds to soil structure and fertility. The deployment of these remains in the form of stubble, differ among various countries. Some opt to use stubble as an option to feed animals, nutrient added value compost, and mushroom cultivation while others choose to burn it on the fields. China, Indonesia, Nepal, Thailand, Malaysia, Japan, Nigeria and Philippines utilize stubble as source of energy, whereas Philippines, Israel, China use it for composting and Lebanon, Pakistan, Syria, Iraq, Israel, Tanzania, China and African countries use it as feed for animals. Open residue burning is a common practice in Asia and in other countries like China, USA, Philippines, Thailand, Indonesia, Taiwan, Pakistan, Nepal and India (Lohan *et al*, 2017).

## 1.1.	STUBBLE PRODUCTION IN INDIA

After crop harvesting, the left-over plant material including leaves, stalk and roots is known as stubble. It is estimated that India generates around 500 Mt of stubble annually with wide regional variability. The uneven distribution and use of stubble depend on the crops grown, cropping intensity and productivity across the nation. A majority of this stubble is in fact used as fodder and fuel for other domestic and industrial purposes. However, there is still a surplus of 140 Mt out of which 92 Mt is burnt each year. The highest stubble estimate was recorded for Uttar Pradesh (60 Mt). Other high stubble producing states are Punjab (51 Mt) and Maharashtra (46 Mt). Cereal crops (paddy, wheat, maize, millets) contribute about 70 per cent to the total stubble production, of which paddy crop contributes around 34 per cent. It is also reported that around 84 per cent of stubble burning is from paddy-wheat system while remaining 16 per cent is from other types of crop rotations. The excess amount of stubble from wheat, barley and pearl millet is used as animal fodder, whereas stubble of cotton and red gram is used as fuel in households. Mustard husks are chiefly used as fuel in brick kilns. Paddy stubble, which is the most generous agricultural biomass from paddy cultivation, is burnt *in-situ*, which is a common management practice in north Indian states such as Punjab, Haryana as well as Uttar Pradesh. In rest of the states like Gujarat, Maharashtra, Tamil Nadu, Bihar, Assam, West Bengal and Jammu & Kashmir, it is used for thatching of houses in rural areas, fuel for domestic cooking and industry, mulching material, compost making, power generation, biofuels, and in boilers for parboiling paddy (Lohan *et al.*, 2017).

1.2.   REASONS FOR BURNING PADDY STUBBLE

The huge amount of stubble left after harvesting of paddy crop can be put to many uses but most of them add to the expenses to be borne by the farmers. While burning the stubble in the field can have the job done with a single matchstick, there are several other important factors contributing to the burning of the stubble. Thus, in spite of multiple potential uses of paddy stubble, famers burn it for the following reasons:

1.2.1. Scarcity of labour for manual harvesting: In agriculture, there has been a progressive shift from manual and animal power to electrical and mechanical means because manual labour and animal power were not sufficient to cope with the work load. The contribution of agricultural workers to the total workers has decreased from 62.67% to 35.96% between 1970-71 and 2012-13 in the northern states like Punjab and Haryana (Lohan *et al*, 2015). The mechanization has tremendously increased due to the shortage of labour, increasing wage rates during harvesting season and falling number of livestock. The traditional method of harvesting paddy with the help of labour using sickles did not leave any significant amount of stubble as the crop was harvested from very close to the ground. The subsequent land preparation for the next crop could be easily done by ploughing and other operations which did not require any burning of the residual stubble. But the gradual shift to mechanized harvesting which ultimately leaves stubble that is managed by burning to avoid further expenses on its management.

1.2.2. Use of combine harvester with the growth of farm mechanization: Farm mechanization has made significant contribution in enhancing agricultural productivity and cropping intensity in northern India. Timely harvest is of utmost importance because a delayed harvest of crops results in a substantial shortfall in grain count as well as straw. This leads to over ripeness causing shattering of grains and hampering the seed bed preparation and sowing operations for the next wheat crop. Traditional method of harvesting paddy requires about 150-200 man-hours/ha. These constraints were overcome by the introduction of combine harvesters. The

Combine Harvester is a machine exclusively used to harvest and recover grain from standing crop. The combine harvester cuts the crop, feeds the crop to the cylinder, threshes the grain from ear head, separates the grain from straw, cleans the grain and handles the clean grain unit in one operation. It provides solutions for scarcity of labour during peak harvesting season and also assists in achieving target in time, minimizing drudgery, reducing crop losses and improving the quality of paddy. On the contrary, combine harvester tends to leave a huge mass of stubble (up to 9.0 t/ha) in the field after harvest and farmers are not equipped and prepared to manage large mass of stubble left in the field. Thus, a farmer often thinks that it is more economical and easier to burn the stubble in the field to enable early sowing.

1.2.3. Timeliness in operation and clearing of field: The time gap between paddy harvesting and planting of wheat in Northern India is 7-10 and 15-20 days in basmati/scented and coarse grain rice, respectively. The combine harvester cuts the paddy crop to a certain height above the ground, thereby creating a huge amount of stubble after harvesting. This stubble is a nuisance for sowing the subsequent crop and in the absence of economic technologies for paddy stubble management; farmers prefer burning the paddy stubble in the field. Traditionally, burning offers a quick approach to empty the field and facilitates further seed bed preparation for sowing of next successive crop, whereas incorporation or collection is perceived to be too costly.

1.2.4. Control of weeds/pests and short-term availability of nutrients: Burning delivers a swift method to control weeds, insects, diseases and pests, both by eliminating them straight away and even modifies their habitat. It is also believed to improve soil productiveness of agricultural land. Residual ash left behind burning acts as a fertilizer and a rich source of potassium. It raises the short-term availability of specific nutrients (e.g. Phosphorus and Potassium) and lowers soil acidity, but leads to a loss of further nutrients (e.g. Nitrogen and Sulphur) and organic carbon. Stubble Burning effectively excludes pathogens on residues, although it has restricted utility for soil disinfection.

1.2.5. Miscellaneous reasons: Some farmers feel that poor storage facilities for straw and lack of market value of stubble force them to burn the stubble in the fields to get rid of it. Few farmers have also realized that the burning in-situ includes reduced tractor fuel cost on left over stubble incorporation by machines.

1.3.   CONSEQUENCES OF STUBBLE BURNING

Production and consumption activities generate pollution and waste and the environment can absorb pollution/waste up to a certain limit depending upon the assimilative capacity in a particular geographical region. Agriculture is one of the most important production activities and stubble burning generates a significant amount of air pollution. If the burning activities remain confined within the assimilative capacity, the pollution does not create harmful effects. Therefore, in the initial stages when the production and burning activities are limited, pollution caused through these activities is not considered a problem. However, due to technological advancements in the agricultural sector, waste concentration has gone beyond the assimilative capacity of the environmental limit, thereby distorting the balance. Stubble burning has become a major cause of atmospheric pollution in the Northern India during paddy harvesting season. Burning of paddy stubble leads to the following consequences:

1.3.1. Depletion of air quality owing to aerosols and trace gas emission: The gases and aerosols arising due to stubble burning consist of carbonaceous matter which play a vital part in global climate change and may lead to regional increase in the levels of aerosols, acid deposition, tropospheric Ozone and depletion of stratospheric Ozone layer. Stubble burning is a process of uncontrolled combustion through which Carbon dioxide is emitted into the atmosphere along with Carbon monoxide, unburnt Carbon, Nitrogen oxides and Sulphur dioxide. The burning of one tonne of paddy straw liberates 3 kg particulate matter, 60 kg Carbon monoxide, 1460 kg Carbon dioxide, 199 kg ash and 2 kg of Sulphur dioxide. It also emits huge quantity of particulates that comprise of a variety of organic and inorganic species. All of this depletes the air quality and makes it highly unfit for living beings.

1.3.2. Liberation of soot particles and causing smog in the environment: Particulate matter which is generated due to stubble burning, being extremely light weight, can stay in the air for a long time causing smog which travels hundreds of miles along with wind. All emissions from burning stubble are fugitive and smoke outflows through unplanned exit routes in the downwind direction. Smog also causes poor visibility and increases traumatic road accidents and other mishaps. At locations of dense smoke, journey takes approximately 20 per cent more time, impacting on time as well as fuel costs.

1.3.3. Health hazards to humans, animals and birds: The incineration of stubble contributes to emission of harmful air pollutants, which can cause severe impact on human health. These include aggravated chronic heart diseases and lung ailments, besides respiratory problems such as asthma, coughing; particularly affecting children, geriatrics and pregnant women. Various studies have also ascribed greater threat of leukemia, blood bone marrow disease, vertigo, drowsiness, headache, nausea, aplastic anemia, and pancytopenia and myelodysplastic syndrome cytopenia to benzene exposure (Chandra and Sinha, 2016). Total annual welfare loss in values of health damages due to air pollution triggered by the incineration of paddy residue in Punjab state is around Rs. 76 million (Kumar *et al.*, 2015). Inhaling fine particulate matter (FPM) adversely affects animal health too. It causes corneal irritation and temporary blindness and chronic bronchitis leading to asthma like conditions. Severe exposure leads to potential decrease in milk yield, and sometimes death of animals. Farmer friendly pests and microorganisms like bacteria, earthworm, etc. die due to fire. Reptiles like snakes, frogs, earthworms and lizards die in the holes. Reports suggest that sparrows, eagles, vultures are becoming extinct possibly due to stubble burning. Nests of birds are also shattered due to this practice.

1.3.4. Deterioration in soil health and fertility: Heat from burning of stubble raises soil temperature leading to death of bacterial and fungal population. However, the death is temporary as the microbes regenerate after a few days. Repetitive burning in the field, however, permanently diminishes the microbial population. Stubble burning kills micro flora and fauna beneficial to soil, thereby depleting the organic matter in the fields. It has been estimated that about 23-73 per cent of nitrogen is lost and the fungal and bacterial population decreases immediately. The burning of stubble raises the temperature of the soil to such an extent that the carbon-nitrogen equilibrium in soil also changes rapidly (Singh *et al.*, 2010).

1.3.5. Loss of plant nutrients/ vegetation: Stubble burning affects nutrient budget and leads to resource loss and harms soil properties. Carbon, Nitrogen and Sulphur present in the stubble are entirely burnt and lost to the atmosphere. The retained stubble enriches the soil, predominantly with organic Carbon and Nitrogen. These

20,729 cases of stubble burning reported in Punjab till November 1

The Punjab Chief Minister's Office has said that as many as 20,729 cases of stubble burning have been reported in the state till November 1, with action being initiated against 2,923 farmers. The air quality index (AQI) in several parts of Delhi reached 1,000 today, with schools being shut in Noida and Delhi due to pollution till November 5.

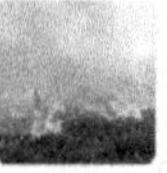

The Indian EXPRESS

Tackling paddy stubble: Super Seeder machine to be launched today

Its main function is to plough standing paddy stubble in soil and sow wheat seed simultaneously in a single operation

The Tribune

Stubble-burning up in Punjab, Haryana

New Delhi: Union Minister of Environment Prakash Javadekar said stubble burning in Punjab and Haryana increased by 75% and 25%. . .

1 week ago

THE HINDU

HOME NEWS OPINION BUSINESS SPORT ENTERTAINMENT CROSSWORD+ SCIENCE

NEWS NATIONAL INTERNATIONAL STATES CITIES

NEWS · NATIONAL

NATIONAL

Punjab witnessed 44.5 % increase in stubble burning incidents in 2020: Centre tells SC

PTI

NEW DELHI, FEBRUARY 02, 2021 13:45 IST
UPDATED: FEBRUARY 02, 2021 19:45 IST

133 farmers penalised for burning stubble in Haryana

The district administration in Haryana's Kaithal has penalised 133 farmers for resorting to stubble burning with the minimum fine of ₹2,500. "The highest fine imposed was ₹15,000. Fines were levied after taking stock of the area of their land," an official said. Notably, Haryana CM Manohar Lal Khattar on Saturday announced ₹1,000 cash reward for giving information on stubble burning.

"FIR against those who fail to pay the fines"
Tap to read what more the official said

India Today

Smog in India could continue until spring, threaten lifespan of many: Report

... cooking, heating and crop stubble burning which are able to accumulate over the region due to topography and cold stagnant conditions.".

2 days ago

Plate 1 : Stubble Burning- the 'burning' topic in the news columns

nutrients then have to be replenished through organic or inorganic fertilizers, which come at a cost. In addition to complete amount of Carbon, 80 per cent of Nitrogen, 25 per cent of Phosphorus, 50 per cent of Sulphur and 20 per cent of Potassium existing in stubble is lost due to burning (Chauhan *et al.*, 2012). Plants and trees standing on bunds, road sides and canal sides are adversely affected due to stubble burning. Micro flora and fauna present in the soil are also destroyed due to burning which leads to loss of biodiversity.

1.4.   ALTERNATIVE PADDY STUBBLE MANAGEMENT MEASURES

Ever since the problem of stubble burning has gained the limelight, many possible alternatives to manage the paddy stubble have been suggested by various platforms including government authorities and scientific communities. Many of those alternatives have been explored by the farmers on their own as well as by the pressure from the government authorities. Some of them are already being practiced by farmers such as utilizing the stubble for mushroom cultivation, using Zero Tillers and other subsidized machines to directly sow the subsequent crop without burning the stubble, etc. But still, a lot has to be done to lead farmers to a situation where they do not have to burn the stubble at all or at least reduce its extent. Paddy stubble can be utilized in many ways instead of burning. Some of the alternative uses of paddy stubble are briefly described below:

1.4.1. Retention of stubble on soil surface: The stubble remained after harvesting can be retained on the soil surface instead of burning. The stubble is spread on the field either mechanically or left as such like mulch for subsequent drying and decomposition. The benefits of retention of stubble on soil surface are i) lesser weed growth, ii) reduced weedicide cost, iii) improved physical, chemical and biological attributes of soils, iv) recycling of plant nutrients, v) lesser fertilizer use in the successive crops. Moreover, stubble retention on soil surface helps in soil moisture conservation by reducing evaporation losses and increasing water holding capacity during wheat growing season. It also increases infiltration, reduces formation of soil crust and runoff. Resource Conservation Technologies (RCTs) based farm machinery provides a better promise in managing paddy stubble for improving soil health, productivity, reducing pollution and achieving sustainable agriculture. For direct seeding of successive crop in loose and anchor stubble load up to 10 t/ha, advance technology of zero-till seed-cum-fertilizer drill/seed planters, have been developed. These technologies are incredibly valuable for managing stubble for controlling of weeds, conserving soil moisture content and nutrients. Punjab Agricultural University developed Happy Seeder which consists of a straw managing unit and a sowing unit in one composite machine, which is mounted on tractor. It facilitates simultaneous spreading of leftover stubble and sowing of subsequent wheat crop in a single operation. For uniform spreading of paddy stubble after harvesting by combine harvester, a Straw Management System (SMS) has also been developed by Punjab Agricultural University. The government of Punjab also made it mandatory for combine harvesters to have mounted the SMS over them in 2015-16.

1.4.2. In-situ incorporation: In-situ incorporation involves mixing of the stubble into the soil after harvesting. It enhances decomposition of combine harvested stubble to enhance nutrients in the soil and can be useful. Stubble incorporation in the soil has several positive impacts on soil health attributes such as pH, organic carbon, infiltration rate and water holding capacity. It increases hydraulic conductivity, cation exchange capacity (CEC), and reduces bulk density of soil by modifying soil structure and aggregate stability. It facilitates surface crust formation, water evaporation from the top few inches of soil and prevents leaching of nutrients. It also increases the microbial biomass and enhances activities of enzymes. An increase in organic carbon increases bacterial and fungal populations in the soil. In 2018, a new machine named Super Seeder was introduced in Punjab which resembles to Happy Seeder in performing dual functions in a single operation. But instead of spreading the stubble on the field like Happy Seeder, the Super Seeder incorporates the stubble into the soil. It also facilitates sowing of crop simultaneously like Happy Seeder, thus, performing twin functions at a time.

1.4.3. Collection of residues for off-farm uses: The stubble generated from the paddy-wheat cropping system can be put to many uses, but this is possible if the residue is carried out off the field. In some parts of Northern India, straw reapers are in practice to collect the stubble from the field and it is gaining popularity in wheat stubble collection because of its economical use for feeding animals. For removal and collection of stubble after combine harvesting and using it for off-farm works; straw

Zero tillage

Square shaped bales

Stubble Burning

Super Seeder

Happy Seeder

Plate 2: Various paddy stubble management measures used by farmers in the study area

baler machines are very promising technology and commercially available. These balers, however, recover only 25-30 per cent of potential stubble yield after combining, depending upon height of plant cut by combines. Baler makes rectangular/square or round bales by collecting the loose stubble from the ground. Thus, baler provides a solution for straw management in an environmentally friendly way. Some private factories have come up in Punjab that intake these straw balers for utilizing them as fuels for various purposes. These factories rent out these baler machines at Rs. 1000 per acre to the farmers and clears off their fields making it ready for subsequent sowing of the next crop.

1.4.4. Feed and bedding: Crop residues are fed to domestic animals in forms ranging from traditional stubble-grazing of harvested grain fields to preparation of chopped residue mixes that are made more palatable and nutritious by the addition of nitrogen-rich compounds. The wheat stubble is widely used for feed purpose while paddy stubble is avoided for using as feed material because it contains Silica which is harmful for cattle. Because of their excellent water-absorption capacity, cereal stubble, especially paddy stubble remains preferred material for animal bedding.

1.4.5. Electricity generation: For briquetting, gasification and power generation, power plant running on producer gas consumes biomass (crop stubble) of 1.2-1.5 kg/kW, so crop stubble is a better option for power generation. Some plants have already been established that are currently utilizing paddy stubble for power generation after making them into pellets and briquettes. At present, over 12 lakh MTs of paddy stubble is being utilized for bio-mass and energy-based projects.

1.4.6. Household fuel: Crop stubble can be used as a fuel. The bulkiness and relatively low energy content of stubble make them inferior to wood but they are still an important source of energy in densely populated and arid regions.

1.4.7. Mushroom cultivation: Wheat and rice stubble are excellent substrates for the cultivation of *Agaricus bisporus* (white button mushroom) and *Volvariella volvacea* (paddy straw mushroom).

1.4.8. Composting: Composting is the natural process of rotting or decomposition of organic matter by micro-organisms under controlled conditions. As a rich source of organic matter, compost plays an important role in sustaining soil fertility and thereby helping to achieve sustainable agricultural productivity. The high organic content in paddy stubble makes it an ideal raw material for compost similar to animal manure and food waste. Farmers have been making compost and utilizing them into their farms since a long time.

1.4.9. Miscellaneous: The other uses of paddy stubble include packaging, mat making, straw board, paper making, etc. Building materials like making bricks and walls from straw-clay mixtures is an ancient technique that is still used in the house and shed construction. These are still prevalent in rural areas of Punjab.

1.5.   GOVERNMENT POLICIES ON STUBBLE MANAGEMENT

India is a legislation rich country with reference to pollution. The issue of stubble burning has been debated comprehensively at numerous forums by scientists, engineers and environmentalists. The government officials are also conscious of the harmful consequences of the practice of stubble burning on environment, human and soil health. Eleven major laws exist to control pollution in India and many forums for their implementation in various ways. Under these laws, provisions have been made to protect the environment from all categories of pollution associated with industrial and agricultural activities. However, to prevent the burning of stubble, the government invoked Section 144 of the Civil Procedure Code (CPC) to ban the burning of paddy stubble, but it is hardly implemented, and there is scanty effort to sensitize farmers on this issue. A National Policy for Management of Crop Residue (NPMCR) has also been formulated and disseminated to all the states for implementation. It aims to ensure prevention of stubble burning by providing incentives to the farmers on the purchase of modern machineries to minimize left over stubble in the field, multiple uses of stubble and formulation of fodder pellets as briquettes for gasification.

Some of the laws which are in operation to regulate the pollution viz. (i) Air Prevention and Control of Pollution Act, 1981 (ii) The Environment Protection Act, 1986 (iii The Environment (Protection) Rules, 1986 (iv) The National Environment Tribunal Act, 1995 (v) The National Environment Appellate Authority Act, 1997. Haryana State Pollution Control Board (HSPCB) has also taken various measures to limit the amount of industrial pollution in the state but not much has been done to address agricultural pollution. As per the directions of the National Green Tribunal (NGT), in 2015, suitable coercive and penal action should be taken by the state government against the persistent stubble burning by the farmers, including launching of prosecution.

On the direction of Honourable High Court of Punjab and Haryana regarding imposition of stubble burning, the Government of Punjab imposed mandatory Straw Management System attachment on all the existing and new production of combine harvester. During 2016–17, Haryana Government commenced action against 1406 violators, and recovered a fine of around 14 lakhs from the farmers who continued to disobey orders on burning paddy residues.

In 2018, the Punjab government also launched three mobile applications to combat stubble burning, viz. i) *i-Khet Machine*: for facilitating farmers to have access to the agriculture machinery/equipment for in-situ management of crop residue ii) *e-PEHaL*: for monitoring tree plantation, and iii) *e-Prevent:* to have prompt and accurate information regarding incidents of crop residue burning.

A Project on "Promotion of Agricultural Mechanization for *In-situ* Management of Crop Residue" has been initiated during 2018-19 in Punjab, Haryana and Uttar Pradesh by Indian Council of Agricultural Research in association with Government of India. The project broadly focuses on: (i) Establishing Farm Machinery Banks or Custom Hiring Centers of in-situ crop residue management machinery; (ii) Procurement of Agriculture Machinery and Equipment for in-situ crop residue management (iii) Information, Education and Communication for awareness on in-situ crop residue management. Under this project, a package of machines would be made available to farmer-groups (Self-help Groups/Co-operatives/Societies etc.) on a subsidized basis for the *in-situ* incorporation of the paddy stubble.

Amid rising number of FIRs for burning the stubble, the Punjab and Haryana High Court delivered a ruling in mid-October of 2019 which directed the authorities not to charge penalties mandated by the NGT, on the ground that farmers were distressed economically. This, however, turned out to be the reason for surge in the number of stubble burning incidents.

The Supreme Court of India in November 2019 directed the Punjab government to provide support of Rs. 100 per quintal to small and marginal farmers who had not burned the stubble by then.

The legislative and judiciary bodies of the nation have attempted to curtail this problem through numerous measures and campaigns designed to promote sustainable management methods and avoid the stubble burning. However, the alarming rise of air pollution levels and health hazards caused by stubble burning in northern India observed in recent years, suggest that the issue is not yet under control. According to official figures, the year 2019 featured a 30 per cent increase in the number of stubble burning incidents as compared to 2018 in Punjab. It seems that the problem of stubble burning is hard to be solved by technical or legal interventions alone. The behavioural aspect of the problem has to be considered as well. It becomes imperative to find out what drives the farmers to burn the stubble in spite of having so many alternatives. An Indian farmer, in general, has undergone a tremendous degree of positive transformation over decades. The farmers are no more just passive recipients of information. They have developed as a character with enhanced socio-economic status, capability and ownership of resources; those who could challenge the scientific opinions with their own experiences. They know what could be the consequences of their actions. Thus, they should not be left out of the equation if we have to find a sustainable solution to the stubble burning. So, its high time that their side of the story is addressed and their motivations, beliefs, perceptions and whatever drives their stubble burning behaviour are explored. Once it is known what causes the farmers to perform such behaviour, only then their behaviour could be changed. As such, the focus should turn towards behavioural objectives and behaviour change techniques over knowledge gain and solely technical and informational approaches when planning and implementing anti-stubble burning programs.

Media and communication have been viewed as products and reinforcers of economic growth and development. At the grassroots level, the communication scholars have advocated the use of persuasive marketing campaigns (in areas such as agriculture, population and health) as the most efficient means to transform traditional individuals and societies. Over time, diffusion of innovations research provided a model of communication interventions in local-level projects. Adoption was defined as the process through which the individual arrived at the decision to adopt or reject the innovation from the time he/she first become aware of it. The five stages were: *awareness, interest, evaluation, trial,* and *adoption.* At the *awareness* stage, the recipient was exposed to the innovation but lacked complete information. At the *interest* stage, the recipient sought more information on the innovation. At the *evaluation* stage, the individual in cognition decided whether the innovation was compatible with present and future needs, the values and experiences. At the *trial* stage, the individual made the decision to tryit on a limited scale. At the *adoption* stage, the individual decided to continue full use of innovation. Early diffusion studies also indicated that at the *awareness* stage, the mass media and cosmopolite information sources were influential, while at the *evaluation* and *adoption* stages, interpersonal and localite sources of information were the dominant modes of influence. However, the diffusion model proved increasingly inadequate, as it does not sufficiently account for recipient feedback, which is crucial to the success of communication campaigns. As a result, project planners in both first and third world contexts have increasingly turned to science-based commercial marketing strategies to promote social causes, a process called *social marketing* to guide project communications. Social marketing remains the predominant model used for projects based on large and complex issues funded by bilateral and multilateral aid agencies (Melkote and Steeves, 2015).

1.6. SOCIAL MARKETING: A POTENTIAL SOLUTION

Over a period of time, communication models have undergone various sorts of transformations. From one-way, top-down and source-to-receiver models to the use of opinion leaders, change agents and mass media to transmit persuasive messages, the models of communication have come a long way. The incorporation of social marketing techniques systematically emphasized the challenges of changing the values and knowledge as well as behaviour patterns of the receivers.

Just like extension education, the social marketing also began as the practice and later became a full-fledged discipline. It was first introduced in India during late 1960s to tackle the grave problem of that time, promoting the awareness about HIV/AIDS and the use of condoms (*nirodh*) for safer intercourse. The concept of social marketing was first formally introduced in 1971 and was defined as "the design, implementation, and control of programs calculated to influence the acceptability of social ideas and involving considerations of product, planning, pricing, communication, distribution and marketing research" (Kotler and Zaltman, 1971). Later definitions further extended and clarified its scope as reflected in the following definition: "the application of commercial marketing principles to the analysis, planning, execution and evaluation of programmes designed to influence to voluntary behaviour of target audiences in order to improve their personal welfare and that of their society" (Andreasen, 1995). On synthesizing various definitions, Kotler and Lee (2008) noted a general agreement that, "social marketing is about influencing behvaiour, it utilizes a systematic planning process and applies traditional marketing principles and techniques, and its intent is to bring a positive benefit to the society."

Social marketing is often confused with advertising and social media. But social marketing is much more than just a means of promotion of a message or an idea.It is a full-fledged discipline, which draws its contents and philosophies from various other disciplines such associology, psychology, economics and anthropology in an attempt to fully understand the behaviour of people. The field's preoccupation with questions about its uniqueness or particular character are, perhaps a consequence of this stance, which compares social marketing with other approaches. In this regard, Andreasen (2002) proposed six social marketing benchmarks to test the legitimacy of a social marketing approach, which subsequently were developed by the UK National Social Marketing Centre through the addition of two new benchmarks. The National Social Marketing Centre Benchmarks for Social Marketing are given in Table 1. These benchmarks are presented as an integrated series of concepts, which capture the breadth of the social marketing field and the knowledge on which it is based (Bird, 2010). They represent a strategic view of social marketing which extends beyond a simple communications campaign or a nudge-based intervention, although these elements may also be part of the social marketing tool kit. In this sense, they reflect current thinking among social marketers, that far from being competitors to the field, different behaviour change theories and approaches are all resources on which practitioners can draw. Consequently, these benchmarks have a useful role to play in grounding social marketing and in showing its relationship with other behaviour change approaches

(Dibb, 2014). Once this understanding is gained, it develops products, services and messages which provide people with an exchange they would value.

Table 1: UK National Social Marketing Centre Benchmarks for Social Marketing

S.No.	Benchmark criteria	Description
1	Behaviour change	Intervention seeks to change behaviour and has specific measurable behavioural objectives
2	Audience research/insight	Formative research is conducted to identify target consumer characteristics and needs. Intervention elements are pre-tested with the target group
3	Segmentation	Different segmentation variables are applied so that the strategy is tailored to the target segment
4	Exchange	Consideration is given to what will motivate people to engage voluntarily with the intervention and what benefit (tangible or intangible) will be offered in return
5	Marketing mix	Intervention consists of promotion (communications) plus at least one other marketing 'P' ('product', 'price', 'place'). Other Ps might include 'policy change' or 'people'
6	Competition	Intervention considers the appeal of competing behaviours (including current behaviour)
7	Theory	Using behavioural theories to understand human behaviour and to inform the programmes which are developed
8	Customer orientation	Attaching importance to understanding the customer, their attitudes and beliefs, knowledge, and the social context in which they are placed

Since its inception, the tools of social marketing have been widely used to encourage behavioural changes related to many community issues including recycling, water and energy conservation, childhood obesity, public health initiatives, and waste reduction, to name a few. For creating behaviour change, social marketing has proved as an effective strategy across the globe under various circumstances and thus, can be a potential tool to change the behaviour of farmers regarding stubble burning.

1.6.1. Social Marketing Planning Process

Social marketers take a holistic view of the process by emphasizing the four Ps in the marketing chain: Product, Pricing, Placement and Promotion. For each targeted market group, the appropriate strategy and mix involving four Ps is devised. Drawing on lessons from the diffusion research, social marketers recognize the importance of targeting individuals who have the power to influence others, that is, opinion leaders in areas such as education, health, agriculture, public policy, etc. Each social marketing expert tends to divide the steps in social marketing planning process in various ways. However, ten commonly accepted steps, as mentioned by Kotler and Lee (2008) are:

1.6.1.1. Describe the background, purpose and focus: The social issue to be addressed by the plan includes a statement of the problem. The background factors that led to the emergence of the social issue are summarized. Then a purpose statement that reflects the benefit of a successful campaign and a focus that narrows the scope of the plan's purpose are developed.

1.6.1.2. Conduct a Situation Analysis: Relative to the purpose and focus of the plan, the factors and forces in the internal and external environment that are anticipated to have some impact on planning decisions are described. It is achieved by conducting a SWOT analysis.

1.6.1.3. Select and describe the target audience: The bull's-eye target audience for the marketing efforts is selected and described. A marketing plan ideally focuses on a primary target audience, although additional secondary audiences are often identified and strategies are developed for them as well.

1.6.1.4. Set marketing objectives and goals (behaviour, knowledge, beliefs): Social marketing plans always include a *behavior* objective, something to influence the target audience to do. *Knowledge* objectives include the information or facts the audience needs to be aware of, ones that might make them more likely to perform the desired behaviour. *Belief* objectives relate more to feelings and attitudes.

1.6.1.5. Identify audience barriers, benefits and the competition: *Barriers* are the reasons the target audience cannot (easily) or does not want to adopt the behaviour. *Benefits* are reasons the target audience might be interested in adopting the behaviour or what might motivate them to do so. *Competitors* are the behaviours that target audience prefers over "desirable" behaviours.

1.6.1.6. Write a positioning statement: Positioning is the act of designing the actual and perceived offering in such a way that it lands on and occupies a distinctive place in the minds of the target audience.

1.6.1.7. Develop a Strategic Marketing Mix (Four Ps):

Product: A product is anything that can be offered to an audience to satisfy a want or need. In social marketing, the product may be a desired behaviour, a key perceived benefit for adopting the behaviour, and/or any tangible objects or services that add value.

Price: Price is the cost that the target audience associates with adopting the desired behaviour. Pricing-related strategies shall aim to reduce costs and increase benefits.

Place: Place is where and when the target audience ought to perform the desired behaviour, acquire any related tangible objects, and receive any associated services.

Promotion: Promotions are persuasive communications designed and delivered to inspire the target audience to action. At this step, messages, messengers and communication channels are determined. Major social marketing communication channels include: advertising, public relations, special events, printed materials, special promotional items, signage and displays, personal selling, social media and popular/entertainment media.

1.6.1.8. Determine an Evaluation Plan: An evaluation plan outlines what measures will be used to evaluate the success of efforts and how and when these measurements will be taken.

1.6.1.9. Establish a campaign budget: The price tags for strategies and activities with cost-related implications including product-related costs, price-related costs, place-related costs, promotion-related costs and evaluation-related costs are determined.

1.6.1.10. Outline an Implementation Plan: The implementation plan functions as a concise working document to share and track planned efforts. Most commonly, plans represent a minimum of one-year activities, and ideally two or three years.

A review of the literature reveals much success in using social marketing techniques to encourage behaviour change. However, there is a relatively modest amount of current literature that documents the application of social marketing to extension programs, despite the fact that this approach to programming has been deemed as highly applicable to extension and adult educational efforts (Skelly, 2005). There have been several instances where the use of social marketing has been advocated by various authors. van den Ban and Hawkins (2002) opined that agricultural extension has much to learn from experiences with changing human behaviour in other fields such as health education which heavily relies on social marketing. Rogers (2003) also reckoned that social marketing has been validated as a promising approach to encouraging practice change among target audiences. Yet it has been absent from much of the colloquy about international extension. Warner and Murphrey (2015) also raised the point that consideration for the complexities of human behaviour has been missing from the majority of the extension campaigns related to environmental sustainability and agriculture. Emphasis needs to be placed on changing human behaviour to solve global environmental problems. This emphasis should include a research-based analysis of the target audience's perceptions towards the behaviour; audience segmentation; articulation of specific, measurable behavioural goals; and the use of specific social marketing tools and principles. Adding to it, Warner (2014) identified a synergy between the social marketing approach and extension programme planning as both focus on influencing behaviour as the bottom line and both are successful due to the act of tailoring programming to specific audience needs. It was also pointed out that extension professionals are currently using certain elements of social marketing, and that this approach has great potential. It would be advantageous to explore ways to expand those elements into complete social marketing campaigns.

1.7.   CONCEPTUAL ORIENTATION OF THE STUDY

Any kind of marketing is theory based. It is predicated on theories of behaviour of target audience, which in turn draw upon the social and behavioural sciences (Novelli, 1990). In fact, this is what happens in the practice of social marketing. Social change is an enormous undertaking and "the one with the biggest toolbox wins." The theories spell out the relationships between human behaviours and various individual, social or environmental factors. The factors are sometimes called behavioural determinants. The cause-effect linkages articulated by important theories have been tested through scientific research and have held up fairly well. By laying out concepts in causal sequences, theories help research scholars to map out the causes of problems under consideration. Concepts from behavioural and social science theory can serve as a checklist to make sure that important issues are not overlooked during the course of research. The choice of strategies and channels can be also be informed by theory. Program evaluators can often measure theoretical determinants of a concerned behaviour even when the behaviour itself is unobservable or a behavioural objective is long range. One of many such theories related to behaviour change was given by Ajzen in 1991, which is called as The Theory of Planned Behaviour. It is an extension to his earlier Theory of Reasoned Action (Ajzen, 1985).

1.7.1. The Theory of Planned Behaviour

According to the Theory of Planned Behaviour, the key determinant of human behaviour is the individual's intention to perform a given behaviour (Ajzen, 1991). Thus, people are expected to act in accordance with their intentions. Intentions refer to the motivational factors that influence a behaviour; they represent the willingness and effort people exert, in order to perform the behaviour. This implies that the stronger the intention to engage in a behaviour, the more likely should be its performance. Although the motivation to engage in a given behaviour may be present, its performance, to some extent, depends on non-motivational factors such as availability of opportunities and resources (e.g., time, money, skills, and cooperation of others). These factors represent people's actual control over their behaviour (Ajzen, 1985). The extent to which intention can influence the performance of a given behaviour depends on behavioural controls. In other words, if a person has the required ability (opportunities and resources), and intends (motivated) to perform the behaviour, he or she should succeed in doing so. For instance, the stubble burning practice by the farmers depends on their motivation (intention) and ability (behavioural control) to engage in the behaviour (burning the stubble).

According to this theory, there are three conceptually independent determinants of intention to perform a given behaviour. The first is the individual's **attitude** towards the behaviour, which reflects the degree to which a person has a favourable or unfavourable evaluation or appraisal of the behaviour, which is, in this case, the burning the stubble (personal factor). The second determinant of intention is the social pressure termed as **subjective norm**, which reflects the beliefs an individual hold regarding the expectations of others (concerning the stubble burning) as well as the individual's motivation to comply with these expectations. In other words, it is the individual's perception of what other people will think about him when they see him burning the stubble, and the willingness of the individual to comply with these expectations. The third determinant of intention is **perceived behavioural control,** which refers to people's perception of the ease or difficulty of performing the behaviour of interest (stubble burning) due to availability or lack of opportunities and resources. When people feel they lack the resources or opportunities to burn the stubble, they are unlikely to form strong intentions to perform the behaviour. That is, behavioural control is about the individual's perception of the extent to which internal and external factors may facilitate or hinder his performance of a given behaviour. A person's perception of the availability of these opportunities and resources greatly influence his intention to perform the behaviour in question. Thus, an individual with a stronger behavioural control will have a stronger intention to perform the desired behaviour. As a general rule, the more favourable the attitude and subjective norm with respect to a behaviour, and the greater the perceived behavioural control, the stronger should be an individual's intention to perform the behaviour under consideration.

In order to have a complete understanding of human behaviour and intentions, it is important to explain why people hold certain attitudes, subjective norms, and perceived behavioural control, which in turn determine intentions and actions. According to the theory, behaviour is a function of salient information, or beliefs, relevant to the behaviour in question. It is these salient beliefs that are considered to be the key determinants of a person's intentions and actions. These beliefs have been classified by Ajzen (1991) as behavioural beliefs, which determine or influence people's attitudes toward the behaviour; normative beliefs, which influence the underlying determinants of subjective norms; and control beliefs, which provide the basis for perceptions of behavioural control.

The theory postulates that attitude towards a behaviour is determined by salient beliefs about the likely consequences of the behaviour, and the evaluations of these outcomes. Attitude towards a given behaviour is determined by the individual's evaluation of the consequences associated with the behaviour and by the strength of these associations. For instance, a farmer who believes that performing a given behaviour (burning the stubble) will lead to a save a lot of money will have a positive attitude toward the stubble burning and if he believes that preventing the pollution is more important than his attitude will be negative towards stubble burning. The beliefs that determines a person's attitude toward a given behaviour is termed behavioural beliefs.

Normative beliefs, on the other hand, is when an individual believes that specific individuals or groups think that he should or should not perform the behaviour; and the willingness to comply with these people. Generally, an individual will be under perceived social pressure to perform a given behaviour if he believes that most of the referents with whom he is motivated to comply think that he should perform the behaviour. However, if he believes that the referents think he should not perform the behaviour, he will be under pressure not to burn the stubble.

Control beliefs deals with the individual's belief about the presence or absence of requisite resources and opportunities (such as low cost of burning, high cost of mechanized measures, time saving, etc) that are needed to successfully perform the behaviour; and the strength of each of these beliefs determines behavioural control. These beliefs may be formed based on past experience, information from friends and acquaintances, or by other factors that increase or reduce the perceived difficulty of performing the behaviour in question. Ajzen (1991) argues that if individuals believe that they have more resources and opportunities, and fewer obstacles or barriers, their perceived control over the behaviour will be greater.

In summary, beliefs regarding the consequences or outcomes determine attitudes towards the behaviour; normative beliefs determine subjective norms, and beliefs about resources and opportunities are the underlying factors that determine perceived behavioural control.

By using the Theory of Planned Behaviour, the extent to which the constructs of the theory (attitude, subjective norms and perceived behavioural control) affect the behavioural intention and subsequently, stubble burning behaviour of the farmers can be identified. On the basis of this, social marketing interventions could be developed focusing on the extent of effect of respective constructs on the behaviour.

1.7.2. Concepts based on Theory of Planned Behaviour used in the study

The present study adopted the Theory of Planned Behaviour to explain the behaviour regarding stubble burning by the farmers. The constructs of the theory have been conceptualized as per the behaviour in the following manner:

1.7.2.1. Stubble Burning Behaviour: According to Ajzen (2020), the Theory of Planned Behaviour starts with an explicit definition of the behaviour of interest in terms of its target, the action involved, the context in which it occurs, and the time frame. Hence, for the present study, the Stubble Burning Behaviour was defined as "Burning (action) the paddy stubble (target) on field (context) after harvesting and before sowing of the subsequent wheat crop (time frame)."

1.7.2.2. Behavioural Intention regarding Stubble Burning: Intention is defined as the motivation; willingness and the effort individuals make to perform a given behaviour. So, behavioural intention regarding stubble burning refers to the degree of readiness of the farmer to perform the act of paddy stubble burning. According to theory, there is a positive relationship between intention and behaviour. The stronger the intention of the farmer means that the higher the likelihood of stubble burning being practiced.

1.7.2.3. Attitude towards Stubble Burning: Attitude refers to the extent to which an individual has favourable or unfavourable evaluation or appraisal of the behaviour in question based on their evaluative or affective judgment. So, attitude towards stubble burning refers to the overall positive or negative evaluation of performing the stubble burning by the farmers. The theory posits that more favourable an attitude is towards the behaviour in question, the higher the intention to perform the behaviour.

1.7.2.4. Subjective norms associated with Stubble Burning: Subjective norms refer to perceived social pressure to perform the behaviour and the willingness of the individual to comply with the perceived pressures exerted by his/her significant others. So, subjective norms associated with Stubble Burning refer to the overall compliance with the influences of the opinions of significant others to the respondent for the act of paddy stubble burning. If individuals perceive that society expects them to perform a particular behaviour, then they are more likely to do so. Conversely, if social expectations were that people should not perform the behaviour, then the individual would be less likely to take actions to perform the behaviour.

1.7.2.5. Perceived Behavioural Control related to Stubble Burning: Perceived Behavioural Control refers to the perceived ease or difficulty of performing the behaviour, which may come from past experience or anticipated impediments or obstacle. So, for the purpose of study, it will refer to the perceived ease or difficulty of performing the act of paddy stubble burning. The more control the farmer feels in burning the paddy stubble, the more likely it is, that he or she will take steps to burn it.

The influence of these constructs on the behaviour was evaluated using appropriate statistical measures and accordingly, an appropriate social marketing plan was proposed in order to change the behaviour of the farmers. Thus, the conceptual framework of the study is given in Figure 1:

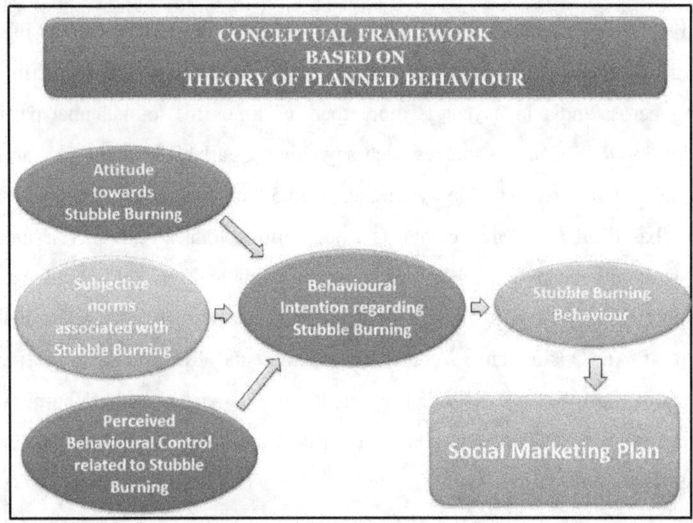

Figure 1: Conceptual Framework of the Study based on the Theory of Planned Behaviour

1.7.3. PROBLEM STATEMENT

Rice-wheat cropping system is the most prominent in India. The Indo-Gangetic plains in Northern India accounts for around twelve million hectares of rice and wheat crop rotation and harvesting of these crops with combine harvesters is very popular with the farmers. The mechanized harvesting leaves a large quantity of stubble that takes a lot of time to decompose in the soil and is usually burnt for speedy plantation of the next crop. India generates around 600 Mt of stubble which includes about 90-140 Mt of surplus stubble and is likely to be burned in the field (Jain *et al.,* 2014*).* Multipurpose use of stubble includes animal feeding, soil mulching, bio-manure, thatching for rural homes and fuel for domestic and industrial use. Despite knowing about alternative methods of stubble management, farmers burn a significant portion of the crop stubble on-farm so that the succeeding crop can be sown on a cleared field. Mechanized farming coupled with lack of availability of farm labour and high cost associated with the process further exacerbates the problem of stubble burning. In fact, according to a study conducted by International Food Policy Research Institute, air pollution due to stubble burning in northern India causes an estimated economic loss of around USD 30 billion annually, and is a leading cause of acute respiratory infections, especially among children; apart from various ill-effects to environment (Chakrabarti *et al.*, 2019). The particulate matter emitted from crop burning across India in a year is more than 17 times the total annual particulate pollution in Delhi from all sources such as vehicles, garbage burning and industries. In Punjab, total cereal stubble generated is 45.58 million tonnes/year and residue burned (based on IARI coefficients) is 21.32 million tonnes/year. In Haryana, total cereal stubble generated is 24.73 million tonnes/year and residue burned (based on IARI coefficients) is 9.18 million tonnes/year (Yadav *et al.*, 2017). In 2019, the System of Air Quality and Weather Forecasting and Research (SAFAR) by the Ministry of Earth Sciences (MoES) reported that the share of stubble burning in air pollution in Delhi-NCR rose to about 46 per cent and thus, public health emergency was declared in the region.

The governments at different levels have attempted to restrict the stubble burning through numerous measures and campaigns designed to promote sustainable management methods. The National Green Tribunal banned stubble burning in 2015 but it had little or no effect in northern states. The charging of monetary penalties from the farmers who burn the stubble and FIRs against them has only politicized the issue rather than putting a brake on it. In spite of the provision of the subsidies for mechanized measures, they are also not reaching the farmers due to bureaucracy and other factors. In 2019, which happened to be the 550th Birth Anniversary of Shri Guru Nanak Dev Ji, the Punjab Government chose to make his holy verse '*Pavan Guru, Pani Pita, Mata Dharat Mahat*' (air is guru, water the father, earth the eminent mother) as the focus of its campaign against stubble burning before the paddy harvesting season began. It was hoped that Guru's eternally-relevant incitement on environment would resonate with farmers more than ever before, as the paddy season overlapped with the 550th 'Parkash Parv' of the first Sikh Guru. But the numbers of stubble burning incidents were more than ever which indicates that tapping just the emotional appeal of the farmers was also not enough.

However, it would be interesting to see the latest direction of the apex court to the Punjab government to provide incentives at the rate of Rs. 100 per quintal to the farmers who do not burn the stubble in the upcoming seasons. As of today, despite enormous efforts and measures, the desirable results to bring change in farmers' behaviour to stop stubble burning is still awaited. It seems that the problem of stubble burning is difficult to be solved by technical or legal interventions alone.Research studies have shown that simply trying to tell people to change, or giving them information and expecting them to act on it, may not work if the determinants of certain behaviours are not considered. There is a need to sensitize the farming community by using interventions which can enable farmers to rethink and change their behaviour and have a new perspective in this regard. In order to understand and change behaviour, scientists rely heavily on appropriate behavioural theories to identify determinants of a particular behaviour. The present study utilized the Theory of Planned Behaviour in the Indian context to the problem of stubble burning in order to understand which behavioural determinants are leading farmers to burn the stubble. Also, knowing why farmers behave the way they do is vital to the success of social marketing interventions which consequently result in behaviour change.

1.8.   RESEARCHABLE QUESTIONS

The present study aims at finding the determinants of stubble burning behaviour of the farmers using the Theory of Planned Behaviour and plan suitable social marketing interventions to inculcate a behaviour change in them. Hence, the study was planned around following researchable questions to explore the problem from different angles:

1. What are the socio-economic, psychological, communicational characteristics of the farmers engaged in stubble burning?

2. What is the extent of farmers' intention to burn stubble?

3. How does attitude, perceived behavioural control and subjective norms affect the behavioural intention of the farmers to burn the stubble?

4. How does the intention to burn stubble influence the stubble burning behaviour of the farmers?

5. Which characteristics of the farmers affect the stubble burning behaviour of the farmers?

6. What type of social marketing interventions can be planned that could encourage the behaviour change in the farmers?

Therefore, keeping in view the above issues the study entitled **"A Study on Stubble Burning Behaviour of Farmers in Punjab"** was planned.

1.9. OBJECTIVES OF THE STUDY

The main purpose of the study was to find how does the Theory of Planned Behaviour constructs (attitude, subjective norms and perceived behavioural control) influence the behavioural intention regarding stubble burning and that intention affect the stubble burning behviour of the farmers. On that basis, a social marketing plan could be chalked out in order to facilitate the behaviour change among the farmers related to stubble burning. Thus, the specific objectives of the study were as follows:

1. To study the socio-personal, economic, communication, psychological and situational characteristics of the farmers

2. To measure the stubble burning behaviour of the farmers

3. To determine the extent to which attitude, subjective norms and perceived behavioural control influence the behavioural intention of the farmers regarding stubble burning

4. To find out the extent to which behavioural intention of the farmers regarding stubble burning influences their stubble burning behaviour

5. To determine the relationship between profile characteristics of the farmers and their stubble burning behaviour

6. To propose a social marketing plan to tackle the stubble burning behaviour of the farmers

1.10. SCOPE AND SIGNIFICANCE OF THE STUDY

Stubble burning has been a problem of grave concern in the recent years especially in northern India. Many technological and legal interventions have been introduced by the government at both central and state levels but this menacing act is far from ending. This study aimed to develop interventions in form of a plan based on social marketing principles that could result in change in the stubble burning behaviour of the farmers in Punjab. The course of the study has included segmentation and profiling of the target farmers. It will serve as a valuable tool for the government functionaries to construct and implement various programmes to stop the stubble burning practice. Also, the social marketing plans that were developed during the study could be replicated in other areas of northern India according to their conditions. The present study will help the extension agencies and policy makers to plan and design similar interventions pertaining to other issues of farmers that pose a potential threat to environment, e.g. ruthless use of chemical fertilizers, etc. So far, social marketing has remained almost absent from the colloquy of extension education, despite being a proven tool for creating behaviour change among different cadres of masses. This study would also pave the path for bringing the two disciplines; i.e. extension education and social marketing to a common platform and harness the benefits of both in creating a sustainable behaviour change process. The findings of the study will serve as a guideline to the extension scholars and personnel to use lesser explored potential tools like social marketing into their projects for crafting a process of behaviour change.

1.11. LIMITATIONS OF THE STUDY

Every care was taken by the investigator to conduct the research systematically. However, some identified limitations during the investigation are given below:

1. Although stubble burning is a common practice in almost all the states of Northern India but the study was carried out in six villages of three districts of Punjab state, so the results and the generalizations may at best be applicable only to the identical village settings.

2. The present study was undertaken as part of the requirement of the Doctoral Degree programme of the student's research. So, there were constraints like availability of time and resources. Therefore, some issues could not be explored in greater depth.

3. Although the tools and techniques for data collection and analysis were developed after going through the sufficient literature and holding consultations with the experts and presented for accuracy and modification, there could always remain a scope for further improvement in the data collection instruments through further refinement.

4. This study largely relied on the responses of respondents and the opinion of the researcher. Therefore, it might have suffered with normal human errors.

1.13.   ORGANIZATION OF THE THESIS

The study conducted has been broadly presented under five heads compiled to form chapters. The first and foremost chapter is named **Introduction,** which includes the conceptual framework, problem statement, researchable questions, objectives of the study, scope and limitations of the study and organization of the thesis. This is followed by chapter two, **Review of Literature**, in which, results of studies already conducted at national and international levels are discussed to provide basic necessary facts about the selected objectives. The third chapter is **Research Methodology**, which presents the procedure adopted for carrying out the research. It describes the sampling techniques, variables selected, construction of tools and techniques used in the study. The fourth chapter is **Results and Discussion,** which highlights the research outcomes and findings and places them in the context of the earlier works through ending discussions. The last chapter is **Summary and Conclusion,** which throws light on the whole research in brief. The literature referred and stated in the study has been acknowledged by enlisting under the head **Literature Cited**. In the end, relevant **Appendices** to the study are provided.

Review
of
Literature

A systematic and thorough review of studies related to the problem under study helps in analyzing the problem, its historical status and development and current status. The reviews of past studies relevant to the present investigation are compiled to develop an understanding on the related dimensions of stubble burning and the application of Theory of Planned Behaviour in agriculture and allied sectors leading to the development of social marketing interventions. The review helps to design the appropriate and useful methodology of the study. Thus, the review of literature has been classified into four subheads:

2.1. Profile characteristics of the farmers

2.2. Stubble Burning and related aspects

2.3. Effect of attitude, subjective norms and perceived behavioural control on behavioural intention and that of behavioural intention on behaviour

2.4. Social marketing interventions

2.1. PROFILE CHARACTERISTICS OF THE FARMERS

Venkatesan (2000) in a study on farmers' awareness and adoption level of recommended tomato cultivation practices in Tamil Nadu stated that nearly half of the tomato growers (48.33%) had medium level of information-seeking behaviour, whereas one-third (33.33%) of them had low information-seeking behaviour. Nearly one-fifth (18.34%) of the respondents had high level of information-seeking behaviour.

Pabhakar (2010) reported in a study on farmers' perception and opinion regarding private extension services in Uttarakhand that 65 per cent of the farmers possessed large land holding and were the major clientele of the private extension services.

Kakkar (2011) in a study on reactions of farmers regarding subsidies in Haryana found that more than half of the farmers (55.83%) had medium level of mass media exposure, 67 per cent had medium level of extension contact and 67.50 per cent of the respondents were the members of the cooperative societies.

Mohapatra (2013) conducted a study on adoption of soil health management (SHM) practices by farmers in Odisha and reported that majority of farmers (53.34%) belonged to medium socio-economic status. Majority of the respondents had medium level of knowledge regarding selected SHM practices (64.44%).

Roy (2015) reported in a comparative study on paddy straw management in Punjab and West Bengal that in case of Punjab, half of the farmers belonged to 21-36 years of age and maximum number of farmers (36.67%) were educated up to senior secondary level. It was also revealed that a little less than half of the farmers (48.33%) had a medium (10-25 acres) land holding while only 5 per cent of the farmers had large (>25 acres) operational land holdings. It was reported that 60 per cent of the farmers in Punjab belonged to the low-income group and 41.67 per cent of the farmers had medium mass media exposure. The study also revealed that half of the farmers had medium level of extension contacts. It was found that farmer-extension linkage was not very strong and farmers' visit to the various agricultural organizations such as State Agricultural Universities (SAU) and Krishi Vigyan Kendra (KVKs) were not very frequent. It was reported that more than 56 per cent of the farmers had a medium level of innovativeness, half of the farmers had a low-risk orientation while 63.33 per cent of the farmers had a high level of ecological consciousness in Punjab.

Verma (2015) in a study on farmers' attitude towards *e-choupal* in Uttar Pradesh found that majority of respondents were middle-aged, educated up to high school, belonged to medium family size, from joint family, small farmers and belonged to middle-income group. The results also revealed that majority of respondents had medium level of mass media ownership, mass media exposure, extension agency contact, information seeking behaviour, social participation, cosmopoliteness and innovativeness. Further, majority of the respondents had medium level of risk orientation, moderate scientific orientation and economic motivation.

Joshi (2016) reported in a study on farmers' perception on climate change in Uttar Pradesh that maximum number of respondents were of young-age (62.5%), illiterate (30.83%), men (98.33%) and had medium family size (78.34%). The study further revealed that maximum number of farmers had low farming experience (85.83%), medium size of land holding (47.50%), low annual income (56.67%), high level of economic motivation (62.5%), medium level of scientific orientation (71.65%), low level of mass media exposure (60%) and medium level of information seeking behaviour (91.66%).

Anand (2016) in a study on stakeholders' opinion on subsidies in Punjab found that a little less than half of the farmers (48.89%) belonged to an age group of 38-49 years, 38.89 per cent were educated up to senior secondary level, 38.89 per cent of the farmers had a medium land holding (10-25 acres), followed by 35.56 per cent having semi-medium (5-10 acres) operational land holding.

Singh *et al.* (2016) conducted a comparative study on socio-economic analysis of the migratory and stationary apiary units in Mansa and Ludhiana districts of Punjab. It was found that majority of the respondents were falling in age group of 31-40 years (56.7% and 60%, respectively). It was also reported that most of the respondents (40% and 53.3%) were having education qualification up to matriculation in both the categories. The respondents having migratory apiary units were mainly small farmers (56.67%) followed by marginal category farmer (20%) and landless labourers (13.33). While 60 per cent, 16.67 per cent and 13.33 per cent of the respondents had stationary apiary units in small farmer, marginal farmers and landless categories respectively.

Deepika (2017) reported in a study on farmers' perception on their vulnerability towards weather variability in Punjab that majority of respondents belonged to old age category, were matric passed, possessed medium-sized land holding and had low annual income. It was also reported that the majority of the farmer respondents possessed medium mass media exposure, extension participation and had poor research and extension linkages.

Parganiha and Sharma (2017) in a study on farmer's perception about climate change in plain zone of Chhattisgarh reported that most of the respondents (47.50%) belonged to middle age group (46-60 years), whereas, 33.75 and 18.75 per cent of them belonged to young age (35-45 years) and old age (more than 60 years) respectively. Majority of the respondents (85.83%) belonged to other backward class, followed by 6.67, 4.17 and 3.33 per cent of general, scheduled caste and scheduled tribe categories respectively. Among all the respondents, maximum proportion was educated up to middle to higher secondary level (35.42%). Most of the respondents (56.67%) were residing as joint family with 5 to 8 members in their family. It was also found that 58.33 per cent of the respondents had 21 to 40 years of farming experience with membership of two social organisations and 57.91 per cent of the respondents participated regularly in organisation like cooperative society.

Sharma *et al.* **(2017)** in a study on farmers' knowledge towards aerobic rice cultivation in Muktsar District of Punjab found that majority of the respondents (62.50%) belonged to middle age group and all respondents were educated with majority (54.17%) of them had passed Intermediate. The farmers predominantly had nuclear family (73.33%) and majority of them (71.67%) possessed 4 to 10 ha of land holding and their annual income ranged between Rs. 2.5 lakhs to Rs. 6 lakhs (65%). Majority of the farmers (65%) had high to medium level of extension contacts, risk taking capacity (55%) and innovativeness to adopt aerobic rice cultivation (58.33%).

Mandal (2018) conducted a study on assessment of knowledge level of farmers regarding soil health management (SHM) practices in West Bengal. The study revealed that the majority of the respondents belonged to middle-aged group (64.17%), had education upto high school (27.5%), belonged to nuclear family (73.33%), with medium level of annual income (86.67%) and had farming as the sole occupation (73.33%). More than half of the respondents (55%) were marginal farmers, with medium farming experience (60%) and had medium cropping intensity (48.33%). It was also found that majority of the respondents (45%) practiced diversified farming, had low social participation (59.17%), medium mass media utilization (59.17%), innovativeness (68.34%), scientific orientation (65.83%) and information seeking behaviour (48.33%).

Palvi *et al.* **(2018)** reported in a study on listening behaviour of farmer audience in Madhya Pradesh that majority (51.67%) of respondents belong to middle age group followed by young age group (38.33%) and old age group (10%) respectively. In case of education, maximum number of respondents (31.66%) had studied up to higher secondary level, followed by middle school (25%), primary school (14.17%), can read and write (12.50%), can read only (7.50%), college level and above (5%) and illiterates (4.17 %). It was also noted that majority of the respondents (65%) belonged to nuclear families followed by joint families (35%). In case of size of land holding, majority of the respondents (50%) were small farmers, followed by medium (34.16 %), large (9.17%) and marginal (6.67%) farmers respectively. It was reported that majority of the respondents (63.34%) had medium annual income followed by low (20%) & high (16.66%) annual income respectively. In case of social participation, majority of the respondents (49.17%) had medium social participation followed by low (46.67%) and high (4.16%) social participation respectively. The extension participation of majority of the respondents (58.33%) was found to be of medium level followed by low (36.67%) and high (5%) level of extension participation respectively. In case of economic motivation, majority of the respondents (55%) had medium economic motivation followed by high (39.17%) & low (5.83%) economic motivation respectively. The majority of the respondents (58.33 %) had medium information seeking behaviour followed by low (40%) and high (1.67 %) information seeking behaviour respectively.

Saklani (2018) found in a study on farmers' perception about extension services of KVKs in Uttarakhand that majority of the respondents were female (55%), middle aged (74.16%) and had secondary level of education (38%). It was also found that majority of the respondents belonged to medium sized family (46.67%), had marginal land holdings (55.83%), belonged to medium income group (66.67%) and had medium level of media exposure (60%). The study further revealed that maximum number of respondents had low social participation (50%), possessed medium level of information seeking behaviour (54.16%), medium level of risk orientation (54.16%), medium level of economic motivation (85%) and medium level of innovativeness (76.67%).

Singh *et al.* (2019) in a study on application of mobile phone agro-advisory services in climate-smart agriculture in Meghalaya found that more than 60 per cent (61.71%) of the respondents belonged to middle age group, followed by 33.88 and 4.41 per cent in the young and old age group. Data also revealed that more than fifty per cent (52.34%) of the respondents had High School level of education which was followed by around one-third (32.78%), one-tenth (11.57%) and 3.31 per cent of farmers who had education up to Higher Secondary, Graduation and above, and were illiterate respectively. It was reported that more than fifty-five per cent (55.92%) had medium sized land holdings, followed by 23.69 per cent and 20.39 per cent of the respondents who had large and small land holding for performing their agricultural and allied enterprises.

Naik *et al.* (2020) in their study in Maharashtra found that majority of farmers were in middle-age groups (52.54%) and males (98.31%). Maximum number of farmers (40.68%) were educated up to graduate level, had progressive farmers (27.71 %) as the main source of information and 55.93 per cent of them were having farm size between 0-2 hectares.

Sharma *et al.* (2020) reported in their study in Punjab that majority of the farmers (51.5%) were above the age of 45 yr. The education level of about one third of the respondents (35.5%) was up to middle school followed by matriculation (29.4%) senior secondary (11.8%), graduation (6.5%) and illiterate (5.5%).

2.2.   STUBBLE BURNING AND RELATED ASPECTS

This section of literature review gives an overview of the field of inquiry that has already been done on the stubble burning and related aspects. In various literatures, the term "crop residue", "crop waste", "crop straw", etc. are used synonymously with the term "crop stubble" referring to the left-over part of the crop plant. For the present study, the term "crop stubble" has been used.

Verma and Bhagat (1992) found that incorporation of rice residue 30 days before sowing of wheat crop resulted in lower wheat yields as compared to the wheat yields when the rice residue is burnt or removed from the fields. Furthermore, the incorporation of rice stubble in the soil has favourable impact on the soil's physical, chemical and biological properties such as pH, organic carbon, water holding capacity and bulk density of the soil.

Gupta *et al.* (2004) in a study on causes and effects of crop residue burning in Punjab estimated that one tonne of straw on burning releases 3 kg particulate matter, 60 kg of CO, 1,460 kg of CO_2, 199 kg of ash and 2 kg of SO_2.

Sidhu and Beri (2005) stated that total production of paddy stubble in Punjab in 2004–2005 reached 18.8 million tonnes, of which 15 million tonnes were burnt in open fields. The study further quoted that 80 % of the rice harvested using combined harvester is burnt in open fields.

Garg (2008) estimated the contribution of rice and wheat stubble loads in the total stubble as 36 and 41 %, respectively in the year 2000, while the contribution of Punjab in the total burnt stubble of rice and wheat to be 11 and 36 %, respectively during the same time period.

Punia *et al.* (2008) attempted to estimate district-wise burnt area of agricultural residue by using remote sensing data. It was found that the total stubble burnt on the area in Punjab was found to be around 4,315.35 km^2 as on 15 May 2005. Among these, Amritsar had 673.99 km^2 of burnt area followed by Jalandhar, Ludhiana, Firozpur and Patiala districts while Roop Nagar had the least burnt area $(41.36\ km^2)$.

Kumar *et al.* (2015) in a study in Punjab found that paddy stubble burning leads to air pollution and several other problems. Irritation in eyes and congestion in the chest were the two major problems faced by the majority of the household members. Respiratory allergy, asthma and bronchial problems were the other smoke related diseases which affected household members in the selected villages. Almost 50 % of the selected households indicated that their health-related problems get aggravated during or shortly after harvest when crop stubble burning is in full swing during the months of October, November and December. In the peak season, affected families had to consult doctor or use some home medicine to get relief from irritation/itching in eyes, breathing problem and similar other smoke related problems. On an average, households spent around more than a thousand Rupees on the non-chronic respiratory diseases like coughing, difficulty in breathing, irregular heartbeat, itching in eyes decreased lung function etc., during the year 2008–2009. However, out of this total expenditure, around 40–50 % was spent during the months of October and November during the time of crop stubble burning. There was an additional cost in terms of household members remaining absent from work due to illness.

Lyngdoh (2018) conducted a study on perception of farmers and extension personnel in Punjab related to paddy stubble burning. It was found that perception of the extension personnel and farmer respondents differ for the statements such as pest and pathogens can be controlled by straw burning, weeds can be controlled by open straw burning and burning of crop stubble decreases the yield of milk in milch animals as well as destroys forest trees. Both respondents agreed that open burning of straw has a negative effect on plant health, air, human health, animal health, biodiversity, vehicular traffic and soil health. The study further revealed that majority of the extension personnel and farmer respondents had medium level of knowledge regarding the economic effect of paddy straw, a high level of knowledge regarding environmental effect and medium level of knowledge regarding pictorial identification for paddy wheat straw management respectively. The findings revealed that the major constraints faced by the extension personnel in disseminating straw management alternatives were inadequate and irregular supply of funds, perceived ineffectiveness of technologies and lack of training facilities. While all the farmer respondents agreed that high transportation of the straw, high labour charges and crop residue interference with tillage and seeding operations were the major constraints in adoption of straw management techniques.

Roy *et al.* (2018) in a study on ways of managing paddy straw in Punjab found that all the sampled farmers burnt paddy straw to clear the field for sowing the subsequent wheat crop. Only 10 per cent of each farmers used paddy straw as animal feed and incorporated it in their fields. None of the farmers were found to use paddy straw for fuel purpose, mulch material, thatching and compost preparation.

Saini *et al.* (2018) in a review on the Stubble Burning in Punjab gave the possible approaches to manage the stubble which include in-situ incorporation, Happy seeder, stubble burry scheme, straw decomposing bacteria and fungi, use of crop residue in bio-thermal power plants, mushroom production, use of stubble as bedding material for cattle, making bio-char and bio-compost etc.

Bali and Rawal (2019) opined in a review that burning of crop residue particularly in Indo-Gangetic plains (IGP) contributes significantly in increasing environmental pollution levels and serious health issues besides deteriorating soil health. Working towards millennium development goals, measures to have sufficient food security, conserving resources, saving environment and dealing with numerous health issues caused by global warming and change in climate are the key issues which has to be addressed very efficiently at large-scale levels.

Chakrabarti *et al.* (2019) studied the risk of acute respiratory infection from crop burning in Northern India. The results suggest that burning of agricultural crop residue to clear fields is a major contributor to air pollution. When rice farmers in north-western India burn their fields, fine particulate matter (PM 2.5) concentrations in Delhi, the highly populated capital city located downwind of burning areas, spike to about 20 times beyond the World Health Organization's threshold for safe air. Living in areas where crop burning is intense (measured using daily satellite imaging data over a 5-month period). It is associated with a three-fold higher risk of acute respiratory infection, one of the leading global causes of lost disability-adjusted life years. Children are particularly susceptible to the health effects of crop burning Solutions to eliminate crop burning exist but require further investments. It was also found that crop-burning abatement would be highly cost-effective and, in northern India, would avert disability-adjusted life years equivalent to US$152.9 billion over a 5-year period. The study gradually revealed that reducing crop burning would benefit human health.

Muzamil *et al.* (2020) provided an engineering intervention to prevent paddy straw burning through in-situ microbial degradation. The study was conducted on designing, developing and evaluating mechanical interface on the basis of physical, engineering and mechanical properties of paddy straw.

This section of this chapter attempted to explore the problem of stubble burning from various angles with the support of past researches done in the context. It was found that various issues related to stubble burning have been researched upon including the pros and cons of burning the stubble on field, amount of nutrient-loss, areas in Punjab with maximum and minimum area under stubble burning, problems faced by people in areas where stubble is burnt, perception and knowledge level of farmers and extension personnel regarding measures to avoid stubble burning, developing engineering interventions to prevent stubble burning through microbial degradation etc. The health loss has also been quantified in monetary terms in a study and all of the studies indicate that the stubble burning should be stopped as soon as possible.

2.3. EFFECT OF ATTITUDE, SUBJECTIVE NORMS AND PERCEIVED BEHAVIOURAL CONTROL ON BEHAVIOURAL INTENTION AND THAT OF BEHAVIOURAL INTENTION ON BEHAVIOUR

Artikov *et al.* (2006) studied the influence of climate forecasts on farmer decisions as planned behaviour in Nebraska, USA using Theory of Plannned

Behaviour. The study quantified the relative importance of attitude, social norm, perceived behavioural control, and financial capability in explaining the influence of climate-conditions information and short-term and long-term forecasts on agronomic, crop insurance, and crop marketing decisions. It was found that attitude, serving as a proxy for the utility gained from the use of such information, had the most profound positive influence on the outcome of all the decisions, followed by norms. The norms in the community, as a proxy for the utility gained from allowing oneself to be influenced by others, played a larger role in agronomic decisions than in insurance or marketing decisions. In addition, the interaction of controllability (accuracy, availability, reliability, timeliness of weather and climate information), self-efficacy (farmer ability and understanding), and general preference for control was shown to be a substantive cause.

Fielding *et al.* **(2008)** in a study incorporated the Theory of Planned Behaviour (TPB) to investigate intentions of students in Queensland University, Australia to engage in environmental activism. Consistent with predictions, environmental group membership and self-identity were positive predictors of intentions. Thus, greater involvement in environmental groups and a stronger sense of the self as an environmental activist were associated with stronger intentions to engage in environmental activism. There was also evidence that self-identity was a stronger predictor of intentions for participants with low rather than high environmental group membership. In accordance with the standard TPB model, participants with more positive attitudes towards and a greater sense of normative support for environmental activism also had greater intentions to engage in the environmental activist behaviour.

Bond *et al.* **(2010)** used the Theory of Planned Behaviour (TPB) to gauge farmers' attitudes, subjective norms and perceived behavioural control towards pesticides in combination with Participatory Rural Appraisal (PRA) tools to adapt an extension program promoting Integrated Pest Management practices in Jharkhand. It was found that farmers had a strong behavioural intention and favourable attitudes, subjective norm and perceived behavioural control to apply pesticide in the following season. It was also recommended to use the theory of planned behaviour in extension related programmes.

Sharifzadeh *et al.* (2012) applied the Theory of Planned Behaviour (TPB) as a theoretical framework to analyze the antecedents of agricultural climate information use behaviour in Iran. The results showed that attitude was positively related to farmers' climate information use in farming decisions. Thus, greater attitude towards use of information in farming decisions was associated with stronger intention to engage in behaviour.

Shin (2014) investigated the antecedents to the behaviour to purchase local food in Oklahoma by using the extended Theory of Planned Behaviour with additional considerations of moral aspects and self-congruity theory. Structural equation modeling was conducted and all hypothesized paths were analyzed. Although hypotheses were constructed based on direct correlations between variables, the study also looked into indirect and total effects on actual local food purchase in order to explain the model more comprehensively. Overall, the purchase of local food was found to be a multifaceted and dynamic decision-making process. In addition to the TPB variables, moral norm and self-congruity were found to influence consumers' local food purchase directly and indirectly, indicating that they were meaningful additions to the TPB model.

Al-Anbari (2016) conducted a research on understanding the constraints to the development of the agricultural sector in Oman by the application of the Theory of Planned Behaviour in order to explore the technology adoption decisions of farmers, specifically with respect to modern irrigation and inorganic fertiliser. It was found that for inorganic fertiliser, it was a combination of the aspects of attitude, subjective norm and perceived behavioural control that limit uptake, whereas for irrigation it was primarily based around perceived behavioural control. The results also suggested that those people who influence adoption of inorganic fertiliser by farmers have similar attitudes to the farmers. The analysis revealed that for farmers, social and cultural factors have significant role in the use of inorganic fertiliser and modern irrigation.

Borges *et al.* (2016) used the Theory of Planned Behaviour to identify key beliefs underlying Brazilian cattle farmers' intention to use improved natural grassland. The TPB hypothesizes that adoption is driven by intention, which in turn is determined by three psychological constructs: attitude, subjective norm, and perceived behavioural control. These three constructs are derived from behavioural, normative and control beliefs, respectively. Results showed that attitude had a strong impact on intention, followed by subjective norm and perceived behavioural control. Results also showed that farmers' intention to use improved natural grassland depends on the extent to which farmers think this innovation allows for increasing the cattle density, the extent to which they think family and cattle traders support them in their decision to adopt and the extent to which farmers think they have sufficient knowledge and access to technical assistance.

Zeweld *et al.* (2016) used Theory of Planned Behaviour to analyse the smallholder farmers' intentions towards two practices: minimum tillage and row planting in Ethopia. The findings revealed that attitudes and normative issues positively explain farmers' intentions to adopt both practices. Perceived control also had a positive significant effect on the intention to apply minimum tillage. It was found that when the intention was formed, farmers were expected to carry out their intention when opportunities arise. The factors like perceived usefulness, social capital, and perceived ease of operation were also significant predictors of farmers' attitudes. Further, it was reported that, social capital and training were factors that positively affected the normative issue, which in turn also positively mediated the relationship between training, social capital and intention. Finally, it was concluded that neither the perceived resources nor information from the media significantly affected farmers' intentions. It was also confirmed that social capital, personal efficacy, training and perceived usefulness played significant roles in the decision to adopt sustainable practices. In addition to it, willingness to adopt seemed to be limited by negative attitudes and by weak normative issues. Therefore, the study posited that attention should be given to socio-psychological issues to improve adoption of sustainable practices by smallholder farmers. This could lead to improvements in farm productivity and enhance the livelihoods of smallholders.

Higuchi *et al.* (2017) used the framework provided by Theory of Planned Behaviour to identify to what extent attitudes, subjective norms, past experience and health involvement determine the intention and frequency of fish consumption. From a set of likert scale indicators, a structural model was specified to evaluate the relationships given by the theoretical framework of the TPB. The results showed that the intention to eat fish was determined by personal attitudes, norms and past experience, and intention itself causes the frequency of fish consumption. It was reported that, although consumers' interest in healthy eating was shown to positively influence fish consumption behaviour by theory, Metropolitan Lima fish consumers seem to be not concerned by positive health attributes related to fish consumption. These results were found to have important implications on production decisions, sales and marketing for the promotion of fish in Lima as a means of economic development.

Issa and Hamm (2017) studied the application of Theory of Planned Behaviour on the adoption of organic farming among Syrian farmers. The results of the study indicated that farmers' positive attitudes and perceived behavioural control towards the conversion to organic fresh fruits and vegetables (FFV) production had a positive influence on the behavioural intention to convert. These findings were found to be in line with the TPB assumptions. Subjective norms towards a conversion to organic FFV production, however, played a small role in forming behavioural intention in our study. Further, it was reported that the direct components of behavioural intention (Attitude, Perceived Behavioural Control and Social Norms) were also well predicted by their respective salient beliefs. The behavioural intention was found as the primary predictor of the decision to adopt organic FFV production, followed by the perceived behavioural control towards this behaviour.

Maichum et al. (2017) developed an extended Theory of Planned Behaviour (TPB) research model that incorporated knowledge to investigate consumers' consumption intention and behaviour towards organic food. They derived and examined the model through structural equation modeling (SEM) on a sample of 412 respondents in Thailand. The findings indicated that consumer attitude and perceived behavioural control significantly predicted consumption intention whereas subjective norm did not. Hence, consumption intention had a positive influence on organic food consumption behaviour. Furthermore, the results suggested that TPB model mediated the relationship between organic knowledge and consumption behaviour.

Pino et al. (2017) developed a model based on Theory of Planned Behaviour to determine the intention of Italian farmers to adopt water saving measures in terms of their propensity to adopt innovations and their water footprints. It was found that favorable attitudes towards water saving measures, and the orientations of environmental associations and public bodies favorably influence farmers' intentions to adopt water saving measures. Farmers' innovativeness and water footprints also exert a significant influence on their adoption intentions.

Sukhmani and Gupta (2017) made an effort to fit the model represented by Theory of Planned Behaviour on the behaviour of agricultural waste processors in Punjab. They defined the behaviour as starting the business or growing the business further. The results showed that the attitude has a predominant role in explaining the behaviour. Also, intention played a statistically significant role as mediating variable.

Bardhan (2018) applied Theory of Planned Behaviour for predicting the organic food consumption behaviour of urban consumers in Eastern India. It was found that attitude was the most influencing factor leading to behavioural intentions in consuming organic food over subjective norms and perceived behavioural control. The Model of Theory of Planned Behaviour showed a good fit in the context of organic food consumption.

Gallagher (2018) investigated the attitudes of small-scale farmers towards conservation-oriented farming as prescribed through the Entry-Level Stewardship (ELS) Agri-Environmental Scheme. The study aimed to understand whether values and identity play a role in how agricultural practices are perceived by farmers and whether this affects their behavioural intentions. The Theory of Planned Behaviour was used to find out whether the policy transformations could translate into changing attitudinal and value positions. It was found that the farmers showed pragmatic attitudes towards farm management, in line with traditional productivist agriculture based on economic motivations but awareness of environmental externalities were present within their behavioural intentions to some extent. Farmers in general presented positive attitudes towards ELS participation.

Taghdisi *et al.* (2018) investigated the factors influencing the farmers' intention to use pesticides based on the Theory of Planned Behaviour (TPB) in Iranian province. The results indicated that TPB constructs along with the knowledge variable could predict the pesticide use behaviour intention among the farmers.

Valizadeh *et al.* (2018) performed the ethical analysis of farmers' water conservation behaviour in West Azerbaijan Province in Iran. The results showed that water conservation intention and perceived behavioural control had positive and significant correlation with water conservation behaviour. Furthermore, the moral norm of water conservation, subjective norm of water conservation, and perceived behavioural control had also positive and significant correlation with water conservation intention. It was also concluded that Theory of Planned Behaviour can be applied to analyze farmers' moral behaviour of the water conservation. Using the proposed theoretical framework resulted in the development and extension of existing understanding, regarding complex interactions of socio-psychological variables of water resources conservation.

Wang *et al.* (2018) investigated farmers' intentions in China to comply with pesticide application standards based on an extended Theory of Planned Behaviour. It was examined how perceived behavioural control, behavioural goal, behavioural attitude and subjective norms influenced farmers' intention to comply with pesticide application standards. The results showed that perceived behavioural control, behavioural goal, behavioural attitude and social norms had positive impacts on farmers' intention in abiding by the standards. Among them in determining farmers' intention towards compliance with pesticide application standards, farmers' perceived behavioural control was found to be the most influential factor, while social norms was the least influential factor.

Hall *et al.* (2019) used the Theory of Planned Behaviour to understand Tasmanian dairy farmers' engagement with extension activities to insure future delivery services. The results indicated that attitude towards extension were consistently positive across all the farmer sub-groups. It was found that there was a negative effect of social influence on experienced farmers' intention to re-engage with extension, due to the belief extension activities were targeted to less experienced, younger farmers. Perceived control factors limiting engagement included lack of confidence about existing knowledge, resulting in farmers perceiving extension activities as confronting.

Sok *et al.* (2020) conducted a critical review of research on Theory of Planned Behaviour (TPB) applied in studying farmers' behaviours as reasoned actions. It was reported that in the 124 articles selected for review, the TPB was applied across a broad range of farm management aspects, mostly relating to 'land and landscape' and 'biosecurity and disease control' issues.

The above section of the literature review shows that the Theory of Planned Behaviour has been used in analysing behaviours of farmers and farm-based consumers related to multiple aspects of agriculture in India and abroad. **Rose (2018)** conducted a Social Science Literature Review on understanding farmers' decision-making behaviour. A total of 171 studies related to farmers' behaviour change were reviewed and it was found that almost half of the studies (49%) used Theory of Planned Behaviour as their theoretical framework. These behaviours have been studied across the globe, which include climate forecasting, environmental activism, pesticide application, local food consumption, adoption of advanced and sustainable agricultural practices like organic farming and water conservation, etc. Some of the behaviours of the farmers like pesticide application are congruent to some extent with the behaviour under the present study, given their negative impact on environment as a whole. Thus, this section of the Review of Literature has been taken as a base support to apply the Theory of Planned Behaviour to the behaviour concerned with the present study, i.e. the stubble burning behaviour of the farmers which would help in identifying the influence of the TPB constructs on the behavioural intention and stubble burning behaviour. This help in planning suitable social marketing interventions.

2.4. SOCIAL MARKETING INTERVENTIONS

Singh (2003) developed various social marketing interventions for promoting environmentally sustainable behaviours in higher institutions of Hissar, Haryana. which included newspaper reporting, captioned photographs, handbills (publicity); cable TV ads, displays (advertising); slogan competition, catalogues (sales promotion); computer presentations and a booklet (personal selling) dealing with the garbage management issues in the campus of the institutes.

Blaxall (2004) analyzed the social marketing interventions used by India's SEWA Bank that targets four types of self-employed poor and illiterate women. Use of *banksathis* (local leaders with good credibility) and "Financial Education Program" (trainings to build the financial planning capacity of its members) was made to foster desired behaviour among women.

TERI (2006) in an analytical study on drip-irrigation technologies for small farmers promoted by International Development Enterprises India (IDEI) found the use of "Profit for Progress" method, slogans in local dialect, advertisements in local television, rickshaw and jeep-based playing jingles through loudspeakers and distribution of leaflets, handbills, wall paintings, boards and banners, demonstrations in *haats* and on farmers' fields to deliver its technologies.

Singhal (2010) in a case study mentioned that Bihar-based social marketing agency *Janani* used Edutainment approach to achieve awareness, attitude and behaviour change objectives for promoting gender equality, small family size, reproductive health and community development by airing a new 52-episode weekly soap opera *Taru* from February 2002 to February 2003 in Hindi-speaking states of Northern India which yielded significant results in the behaviour of the community members.

Sorenson *et al.* (2011) assessed the effect of social marketing incentives on dispositions toward retrofitting and retrofitting behaviour among farmers whose tractors lacked rollover protective structures in New York and Pennsylvania. A quasi-randomized controlled trial was conducted with 391 farm owners in New York and Pennsylvania surveyed before and after exposure to 1 of 3 tractor retrofitting incentive combinations. These combinations were offered in three trial regions; region-1 received rebates; region-2 received rebates, messages, and promotion and was considered the social marketing region; and region-3 received messages and promotion. A fourth region served as a control. The social marketing region generated the greatest increases in readiness to retrofit, intentions to retrofit, and message recall. In addition, post intervention stage of change, intentions, attitudes, subjective norms, and perceived behavioural control levels were higher among farmers who had retrofitted tractors. The results showed that a social marketing approach (financial incentives, tailored messages, and promotion) had the greatest influence on message recall, readiness to retrofit tractors, and intentions to retrofit tractors and that behavioural measures were fairly good predictors of tractor retrofitting behaviours.

Andriamalala *et al.* (2013) used social marketing interventions to foster sustainable behaviour in traditional fishing communities of southwest Madagascar. A number of marketing materials carrying the messages were created, including Tshirts, posters, radio broadcasts and songs. Each was designed specifically for the different target audiences to address their individual barriers to change.

DeWan *et al.* **(2013)** studied a social marketing campaign in Gansu province of China aimed to influence household adoption of fuel-efficient cook stoves and conservation of forest habitats for the Sichuan golden snub-nosed monkey. Pre and post-survey data directly measured the change in behaviour after both, a one-year and 2.5-year period. Local campaign managers used the data from pre-campaign surveys as well as site visits to better understand the knowledge, attitudes, interpersonal communication, and barriers for the target audience as well as to segment communities who would be most receptive to a campaign and to which types of messaging. They discovered their target audience was adult community residents and that fuel-efficient stoves can save them time in cutting wood and cleaning the kitchen as well as increase health benefits from cleaner indoor air quality. Through decreasing fuel wood consumption, the campaign further showed how adults would decrease pressure on the forest habitat for the golden monkey. Survey data revealed the following barriers: (1) cost of the stoves; (2) knowledge of how to make, use, and maintain them; and (3) development of a sustainable forest management plan. The campaign sought to overcome these obstacles through providing subsidised stove prices, training, and advisory support for the forest management plan. Campaign managers identified the local market as the ideal place for disseminating materials; they employed a combination of posters, calendars, and a tele-film to distribute the campaign's message and incorporated culturally and socially relevant elements from the community. The use of TV as a promotion tactic emerged as another key insight from the audience research process. Post-campaign surveys showed an increase of 28 per cent and 43.1 per cent adoption, after 1 and 2.5 years, respectively.

Pounds *et al.* **(2014)** made the formal use of social marketing concepts in a systematic approach to influence farmers in USA to voluntarily increase respiratory protective device (RPD) use. The planning process for the project incorporated six key decision or action points, each informed by formative research or health behaviour theory. The planning process included developing behaviour change strategies based on a 4P model (product, price, place, and promotion). The resulting campaign elements included print and e-mail messages that leveraged motivators related to family and health in order to increase farmers' knowledge about the risks of exposure to dusty environments, four instructional videos related to the primary barriers identified in using RPDs, and a brightly colored storage bag to address barriers to using RPDs related to mask storage. Campaign implementation included a series of e-mails using a bulk e-mail subscription service, use of social media in the form of posting instructional videos on a YouTube channel, and in-person interactions with members of the target audience at farm shows throughout the Central States Center for Agricultural Safety and Health seven-state region. Evaluation of the e-mail campaigns indicated increased knowledge about RPD use and intent to use RPDs in dusty conditions. YouTube analytic data indicated continuing exposure of the instructional videos beyond the life of the campaign. The project demonstrates the efficacy of a planning process that incorporates formative research and clear decision points throughout. This process could be used to plan health behaviour change interventions to address other agriculture-related health and safety issues.

DeWitt (2017) organized a social marketing campaign at farmers' markets to encourage fruit and vegetable purchases in rural counties. This study used the "Plate It Up Kentucky Proud" (PIUKP) social marketing campaign in six rural communities over two years to determine the association between exposure to the campaign and fruit and vegetable purchasing decisions. The campaign included distribution of recipe cards, recipe samples of various dishes and special PIUKP food sample. It was found that the availability of recipe cards at the farmers' market was associated with influencing the purchase of fruits and vegetables; the recipe sample was associated with buying the ingredients for the recipe; taking the PIUKP food sample was associated with a willingness to prepare the food item at home. It was recommended that utilizing social marketing strategies at farmers' markets may be an effective way to improve fruit and vegetable intakes in rural communities.

Bazhan *et al.* (2018) applied social marketing mix to identify consumer preference towards dairy products in Iran. The preferences toward functional dairy products were categorized in four main groups: characteristics of products including sensory and non-sensory characteristics; price; place of the product supply; and promotion strategies of products categorized in three subgroups of informing and educating, advertising and recommending.

Gazali *et al.* (2018) tested social marketing strategy on smallholder farmers' flood preparedness behaviour in East Coast Malaysia. Social marketing strategy constructs (core product, barrier identification, media channel and their communication and promotion) significantly predicted flood preparedness behaviour. The study further revealed that in Malaysia, promotion is the strongest predictor of flood preparedness behaviour. Hence, should be included in government policies and programs aimed at influencing behaviour among respondents. Thus, the findings of the study served as an important benchmark in providing information to the government and agencies particularly the Ministry of Agriculture Malaysia, in their quest to design policies and programs to influence farmer's behaviour to adopt flood preparedness.

Sneed *et al.* (2018) examined the impact of a farmers' market social marketing campaign on limited-resource consumers' intent to purchase and consume locally-grown produce and to examine potential differences in campaign impacts based on consumers' communities of residence (rural, suburban, urban) in England. A farmers' market social marketing campaign was implemented in 33 counties at 39 farmers' markets. The campaign consisted of marketing assets, six food demonstrations with recipes, and three children's activities. Participant surveys were conducted with a convenience sample of 15 farmers' markets located in 10 counties (three rural, four suburban, and three urban). Surveys were completed by 723 adults. Most participants reported purchasing and consuming more locally-grown produce as a result of the campaign. There were differences, however, between rural, suburban, and urban consumers' responses with suburban consumers more likely to report increases in purchasing and consuming locally-grown produce. While effective in increasing purchase intention and consumption of locally-grown produce with all consumers, there were differences in campaign impact among rural, suburban, and urban consumers. These findings suggest different social marketing approaches might be warranted when engaging rural, suburban, and urban limited-resource farmers' market consumers.

Dessart *et al.* (2019) in a policy-oriented review on behavioural factors affecting the adoption of sustainable farming practices pointed out that social marketing programmes, including agricultural education and training through advisory services and media campaigns, can, in the long term, raise farmers' environmental concerns and increase the importance they give to conservation as a farming objective. Government-run campaigns are already used to induce farmer behavioural change, for instance to promote more appropriate use of pesticides and chemicals. However, rather than traditional, isolated top-down media campaigns, raising environmental concerns requires building long-term, integrated and diversified social marketing programmes involving all stakeholders, therefore fully including farmers in the process.

Sharma *et al.* (2019) conducted an action research by applying a seven-step approach of social marketing to bring change in water use behaviour of rural people of Punjab, India. Findings revealed that the overall level of awareness of the majority of the respondents regarding climate change was found to be medium. They also suggested that awareness of effects of climate change on water resources and factors responsible for water depletion should be created among farming families to make them water saving conscious. Thus, an awareness campaign consisting of rallies and street plays was conducted. Knowledge regarding water saving technologies was imparted through skill training. Adoption of small water saving devices on a trial basis was facilitated by distributing water literacy kits. Post-knowledge test data revealed that the knowledge of 68 per cent farmers and 61.33 per cent of farm women increased to a high level after intervention. All the respondents adopted distributed technologies on a trial basis. The social marketing approach proved to bring about change in the water use behaviour of rural people and impacted their knowledge and adoption level.

The above section of Review of Literature underpins several examples of social marketing interventions and campaigns which brought a significant behaviour change among diverse cadres of clientele of agricultural and rural communities. The issues or behaviours under concern varied from health and safety of farmers to judicious use of water by village communities. The review also suggests that the key to successful social marketing campaign in bringing the desirable behaviour change is to segment the audience as well as possible on the basis of homogenous characteristics and design interventions specifically tailored according to them. It has also been noted that sky is the limit in the ways of designing the suitable messages and promoting them. Various forms of media can be utilized to cater to the needs of the target audience so that the message could be delivered in the form desired by target audience. It was also noted that social marketing is no one-man's job. It requires multiple stakeholders to work in synergy including government, non-government, private agencies, etc. to bring the desirable change among the target audience.

Research Methodology

The present chapter describes the scientific and systematic procedure followed in the light of the objectives set forth the study. It also described different components of the methodology which are covered under the following heads:

3.1. Universe and locale of the study

3.2. Sample and sampling procedure

3.3. Description of villages

3.4. Research Design

3.5. Selection of variables and their measurement

3.6. Operationalization of variables

3.7. Hypotheses set for the study

3.8. Tools and techniques of data collection

3.9. Type of documentation

3.10. Statistical analysis and interpretation of data

3.1. UNIVERSE AND LOCALE OF THE STUDY

3.1.1. Universe: Punjab- The land of five rivers

Punjab is one of the most prosperous states in Northwest region of India. The name 'Punjab' is made of two words *Panj* (Five) and *Aab* (Water) i.e. land of five rivers. These five rivers are Sutlej, Beas, Ravi, Chenab, and Jhelum. Only Sutlej, Ravi and Beas rivers flow in today's Punjab. The other two rivers are now in the state of Punjab, situated in Pakistan.

The Indian State of Punjab was created in 1947, when the partition of India split the former Raj province of Punjab between India and Pakistan. Most of the western part which consisted of mostly Muslim population became Pakistan's Punjab province; while the eastern part which constituted mostly of Sikh population became India's Punjab state. Several small Punjabi princely states, including Patiala, also became part of Indian Punjab. In 1950, two separate states were created; Punjab included the former Raj province of Punjab, while the princely states of Patiala, Nabha, Jind, Kapurthala, Malerkotla, Faridkot and Kalsia were combined into a new state, the Patiala and East Punjab States Union (PEPSU). Himachal Pradesh was created as a Union Territory from several princely states and Kangra District. In 1956, PEPSU was merged into Punjab state, and several northern districts of Punjab in the Himalayas were added to Himachal Pradesh.

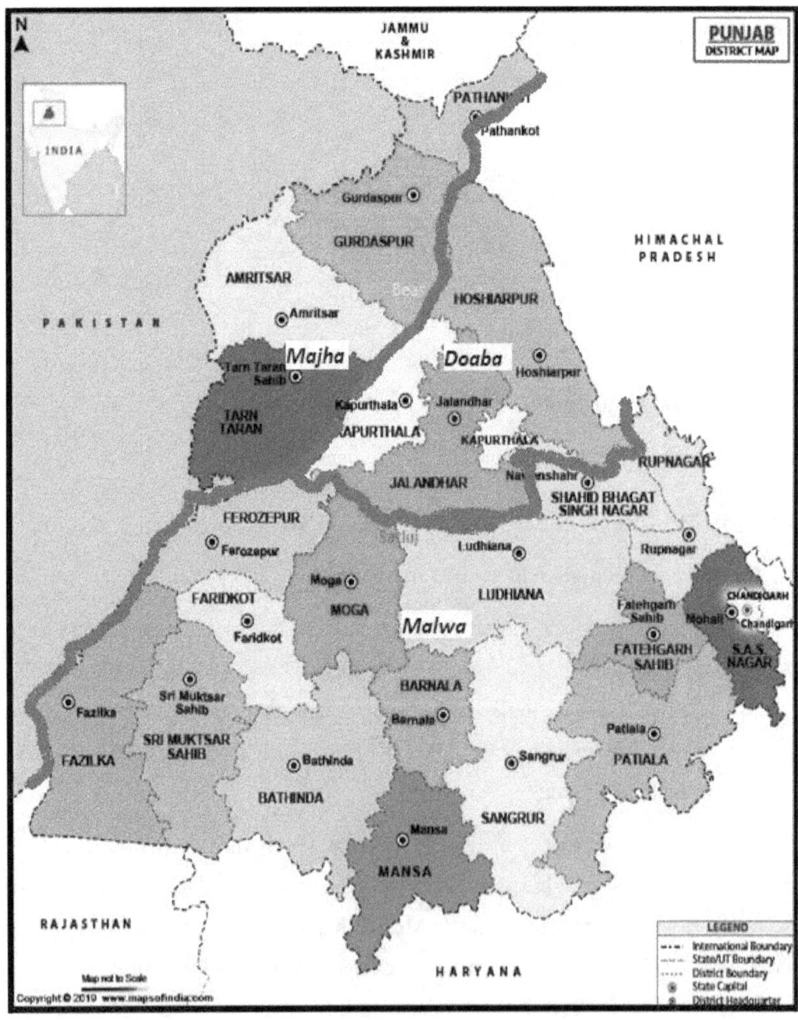

Figure 2: Map of Punjab state (Source: *www.mapsofindia.com*)

3.1.1.1. Geo-ecological description

The total area of the Punjab state is 50,362 sq. km. (19,445 square miles), with most of the cultivable area being under irrigation. Its average elevation is 300 meters (980 ft.) above sea level, with a range from 180 meters (590 ft.) in the southwest to more than 500 meters (1,600 ft.) around the northeast border.Punjab extends from the latitudes 29.30° North to 32.32° North and longitudes 73.55° East to 76.50° East. It is bounded on the west by Pakistan, on the north by Jammu and Kashmir, on the northeast by Himachal Pradesh and on the south by Haryana and Rajasthan.

The Punjab State is divided into three major regions namely: Majha, Doaba, and Malwa. This division of Punjab is basically due to the rivers Sutlej and Beas flowing through the land of Punjab. In the historical times, it was not easy to cross the rivers and hence the areas divided by rivers were considered as separate regions. These regions were often ruled by different rulers or kings. The interaction between the people living in these geographically separated areas was limited. Due to this, there exists a difference between the language and culture of the people living in these regions. The three regions span over the parts of historic Punjab region which includes today's Punjab under Indian boundaries, Haryana, Himachal Pradesh and Punjab now under Pakistan.

The region of Punjab towards the left bank of river Sutlej is called Malwa. Malwa extends upto Ambala district in Haryana, beyond the borders of Punjab state. Almost 60-70 per cent area of Punjab state is part of this region. Modern-day districts of Barnala, Bathinda, Fatehgarh Sahib, Faridkot, Fazilka, Firozpur, Ludhiana, Mansa, Moga, Mohali, Muktsar, Patiala, Ropar and Sangrur lies in Malwa region of Punjab.

The area between the rivers Satluj and Beas is called as Doaba. The word 'Doaba' is made up of two words '*Do*' meaning two and '*Aab*' means water or river, so the literal meaning of word Doaba is the area between two rivers. The districts of Jalandhar, Kapurthala, Hoshiarpur and Nawanshahr (ShaheedBhagat Singh Nagar) fall in this region.

Majha region mainly covers the area between Beas and Ravi rivers. The area on the north of Sutlej, after the confluence of Beas and Sutlej at Harike in Tarn Taran district, extending upto river Ravi is also part of the Majha region. The literal meaning of word 'Majha' means 'in the middle' or 'at the centre'. This area was the central part of the historic Punjab region, hence giving it the name Majha. This area includes a major region of Pakistan's Punjab, extending upto river Jehlum. The Majha region constitutes the modern-day districts of Amritsar, Gurdaspur, Pathankot and Tarn Taran from India.

The state has a balanced amalgamation of heat in summer, rain in monsoon and cold in winter. It experiences both summer and winter to its extreme and even receives abundant rainfall, which makes the state a very fertile land. The region lying near the foot hills of Himalayas receives heavy rainfall while the region lying distant from the hills receives scanty rainfall and the temperature is high. The summer months span from mid April to the end of June. The rainy season in Punjab is from early July to end of September. October marks the beginning of the winter season. From December onwards, the winter becomes chilly which goes on till early February. The spring follows the freezing weather from February and the temperature gradually starts rising which goes on through out summer.

3.1.1.2. Administrative description

The Census of India, 2011 states that Punjab is officially divided among five divisions: Patiala, Rupnagar, Jalandhar, Faridkot and Firozpur. It has 22 districts subdivided into 91 Tehsils, which are further divided into 81 Sub-Tehsils. The state has 150 Blocks, 143 Towns and 74 Cities. The blocks consist of revenue villages and the total number of villages in the state is 12,581. Apart from these there are 22 Zila Parishads, 167 Municipal Committees and 28 Improvement Trusts looking after 143 towns and 74 cities of Punjab. In 1966, Chandigarh city assumed the unique distinction of being the capital city of both, Punjab and Haryana states, while it itself was declared as a Union Territory.

3.1.1.3. Demographic description

As per Census of India (2011), the total population of Punjab is 2,77,43,338, of which 1,46,39,465 are males and 1,31,03,875 are females. The population of Punjab formed 2.29 per cent of total population of India in 2011. The literacy rate in Punjab was recorded to be 76.68 per cent and of that, male literacy rate stood at 81.48 per cent while female literacy was at 71.34 per cent. With total geographical area of Punjab at

50,362 square kilometers, the population density of Punjab was 550 per sq. km. which is higher than the national average of 382 per sq. km. The Sex Ratio of Punjab is 893 i.e. for each 1000 males, there are 893 females in the state; which is lower than the national average of 940. The share of rural population in the total population of the state was 62.51 per cent whereas the proportion of urban population was recorded to be 37.49 per cent. The rural literacy rate has been 72.45 per cent and that of urban literacy was 83.70 per cent as per the latest census data (Census of India, 2011). The detailed demographic description of the state has been further provided in the Table 1.

3.1.1.4. Economic description

Punjab has made considerable economic progress after Independence despite the setback it suffered in 1947. The state has a GDP of 5.78 lakh crore which is approximately US$ 84 billion (Punjab Economic Survey, 2019-20). As one of the most fertile regions in the country, agriculture is the mainstay of the state's economy. The region is ideal for the cultivaton of wheat while the wheat-rice crop cycle is most commonly followed practices in the state. It contributes nearly two-third to the total production of food grains and a third of milk production in the country. The initiative of Green revolution has been keenly taken forward by the people of Punjab. The state now contributes to 10.26 per cent of cotton, 19.5 per cent of wheat, and 11 per cent of India's rice production. At global level, Punjab alone produces two per cent of the world's cotton, two per cent of wheat and one per cent of world's total rice production. The largest cultivated crop in the state is wheat. Other important crops are rice, cotton, sugarcane, pearl millet, maize, barley and fruits (Punjab Economic Survey, 2019-20). Thus, Punjab is called the "Granary of India" or "India's bread-basket".

The state has been awarded the National Productivity Award for Agriculture Extension services for ten years, from 1991-92 to 1998-99 and from 2001 to 2003-04. In recent years, a drop in productivity has been observed, mainly due to degrading fertility of the soil. This is believed to be due to excessive use of fertilisers and pesticides over the years. The consumption of fertiliser per hectare is alarmimg at 223.46 kg as compared to the national figures of 90 kg. Another matter of concern is the rapidly falling water table on which almost 90 per cent of the agriculture depends. By some estimates, groundwater is falling by a meter or more per year.

Other major industries in Punjab include manufacturing of scientific instruments, electrical goods, financial services, machine tools, textiles, sewing machines etc. Even though the population of Punjab accounts for less than 2.5 per cent of the Indian population, the per capita income is twice the national average. According to the India State Hunger Index, Punjab has the lowest level of hunger in India.

3.1.1.5. Cultural description

Punjab has one of the oldest cultures of the world. Its diversity and uniqueness is evident in the Punjabi poetry, philosophy, spirituality, education, artistry, music, cuisine, science, technology, military warfare, architecture, traditions, values and history.The official language of the state is Punjabi which is the tenth most widely spoken language in the world. It is also the fourth most spoken language in Asia. It is the only living language among the Indo-European languages which is a fully tonal language. Punjabi is written in the Gurmukhi Script. Besides Punjabi, Hindi, Urdu and the universally acclaimed English are the languages that are spoken in Punjab.

The people of Punjab celebrate numerous religious and seasonal festivals, such as Lohri, Baisakhi, Dussehra, Diwali etc. There are numerous anniversary celebrations as well in honour of the Gurus (the ten religious leaders of Sikhism) and various saints. Expressing happiness and gaiety through dance is a typical feature of such festivities, with bhangra, jhumar, and sammi being among the most popular genres. Giddha, a native Punjabi tradition, is a humorous song-and-dance genre performed by women.

Poetry offers one of the clearest views into the Punjabi mindset. Punjabi Poetry is renowned for its deep meaning, beautiful, exciting and hopeful use of words. Many compilations of Punjabi poetry and literature are being translated throughout the world into many languages. One of the most important Punjabi literatures is that of the revered "Shri Guru Granth Sahib".

The traditional dress for Punjabi men is 'Punjabi Kurta' and 'Tehmat' or 'Chadra', which is now being replaced by the kurta and pajama in the modern day Punjab. The traditional dress for women is the Punjabi Salwar Suit which replaced the traditional Punjabi Ghagra.

Punjab's geographical location with relation to the rest of the sub-continent has meant that this region has had strong Central Asian influences both in its culture and its food. Punjabi cuisine has become world-leader in the field; so much so that many entrepreneurs who invested in the sector have built large personal fortunes due to its popularity. The "*Sarson ka saag*" and "*Makki di roti*" are examples of well-known and very famous dishes. Punjab's economy has predominantly been agrarian in nature, historically evidenced in the remains of granaries and other artifacts of the Indus Valley Civilization. Dairy products, pulses, vegetable and meat curries continue to reflect the rural temper of the state while being wedded to the residual flavors of foreign invasions, such as rice and gravies. The tradition of *langar* (community kitchen) initiated by Guru Amardas, came from the belief that food is central to communal bonding. It has since been a remarkable feature of all Gurudwaras, wherein devotees of all faiths participate in the preparation and service of meals.

Table 2: Punjab at a glance (*Source: Statstical Abstract of Punjab, 2019*)

S.No.	ITEMS	YEAR	UNIT	STATISTICS
(A)	**Area**			
1.	Geographical Area	2018	km^2	50,362
2.	Rural Area	2018	km^2	48,265
3.	Urban Area	2018	km^2	2,097
(B)	**Population**			
1.	**Total Population**	2011	Lakh	277.43
(i)	Male	2011	Lakh	146.39
(ii)	Female	2011	Lakh	131.03
2.	**Rural Population (Total)**	2011	Lakh	173.43
(i)	Proportion of total population	2011	%	62.52
3.	**Urban Population (Total)**	2011	Lakh	104
(i)	Proportion of total poulation	2011	%	37.48
4.	**Sex Ratio**	2011	No. of females per thousand males	895
5.	**Literacy Rate**	2011	%	75.84
(i)	Male	2011	%	80.44
(ii)	Female	2011	%	70.73
6.	**Population Density**	2011	Persons/ km^2	551
(C)	**Administrative Units**			

1.	Divisions	2018	No.	5
2.	Districts	2018	No.	22
3.	Tehsils	2018	No.	91
4.	Blocks	2018	No.	150
5.	Towns	2018	No.	143
6.	Cities	2018	No.	74
7.	Inhabited villages	2018	No.	12,581
(D)	**Local Bodies**			
1.	ZilaParishads	2018	No.	22
2.	Municipal Committees	2018	No.	167
3.	Improvement Trusts	2018	No.	28
(E)	**Agriculture**			
1.	Net Sown Area	2018	'000 ha	4118
2	Area Sown more than once	2018	'000 ha	3721
3	Cropping Intensity	2018	%	190
(F)	**Irrigation**			
1.	Net area irrigated by:			
(i)	Government canals	2018	'000 ha	1169
(ii)	Wells/Tube wells	2018	'000 ha	2907
(iii)	Total Area Irrigated	2018	'000 ha	4076
2.	Gross Area Irrigated	2018	'000 ha	7758
(G)	**Category-wiseLabour Force**			
1.	Main Workers	2011	Lakh	84.51
2.	Cultivators	2011	Lakh	19.35
3.	Agricultural Labourers	2011	Lakh	15.88
4.	Marginal Workers	2011	Lakh	14.46
5.	Workers in household industries	2011	Lakh	3.86
(H)	**Area, Production and Productivity of major crops**			

S.No.	Crop	Area ('000 ha)	Production ('000 MT)	Productivity (kg/ha)
1.	Wheat	3520	18262	5188
2.	Rice	3103	12822	4132
3.	Cotton	268	1223	1382
4.	Maize	109	396	3625
5.	Sugarcane	95	7774	81828
6.	Rapeseed and Mustard	30.5	46.5	1524

3.1.1.6. Cropping system of Punjab

Punjab is one of the chief contributors to the country's total food grain production. In Punjab, the farmers largely follow the rice-wheat cropping system. Rice is usually grown in the *Kharif* season (sown in July-August and harvested in October-November) and wheat in the *Rabi* season (sown in November-December and harvested in April-May). The major constraint in the rice-wheat cropping system is the availability of short span of time between the harvesting of rice and sowing of wheat.Rice and wheat are double cropped in Punjab with rice stubble being burned off over millions of acres prior to the planting of wheat. Any delay in sowing adversely affects the production and yield of the subsequent wheat crop. The preparation of field involves removal or utilization of rice stubble left in the field.

Over a period of time, various modern inputs have been introduced in Punjab to harvest the rice crop within such a short span of time. One such input is the combined mechanized harvester which has now become the most popular implement used for harvesting in the rice-wheat cropping system. The use of the combined harvester has increased at a tremendous rate in Punjab over a period of time. Almost 80 per cent of the rice crop is harvested by it. However, the use of the combined harvester has also enhanced the problem of crop stubble management. The use of combined harvesters leaves behind a large amount of rice stubble in the fields which is difficult to collect. It has been widely observed that farmers find burning the stubble as the easiest and the most economical way of getting rid of it.The stubble burning also facilitates the farmers to speed up the sowing of subsequent wheat crop in the short span of time. Thus, burning has emerged as the standard method of rice stubble management in the combine harvested rice-wheat cropping system that is practised on a broad scale in Punjab.

Every year, almost 15 million tonnes of paddy stubble is generated in Punjab. Of this, according to various estimates, on an average, almost half of the rice stubble is set on fire in open fields (Kumar *et al.*, 2015). The stubble burning is monitored through Satellite Remote Sensing by NICRA, IARI-CESRA, New Delhi. The data highlights that the total number of stubble burning incidents recorded during 1st October 2018- 30th November 2018 were 75,563. Out of which, majority of the burning incidents were recorded in Punjab (59,695). Whereas the number of stubble burning incidents in Haryana and Uttar Pradesh were found to be 9232 and 6696 respectively (Chhabra *et al.*, 2019). The above-mentioned period is the peak time for paddy harvesting in these states.

3.1.2. Locale of the study: Malwa region

According to Kumar *et al.* (2019), the majority of the stubble burning incidents in North India were recorded in Punjab alone. While in Punjab, the Malwa region accounted for almost 95 per cent of the total crop stubble burning incidents. Therefore, the **Malwa region was selected purposively as the locale for the study**.

The data in Figure 3 shows the district wise fire counts percentage in Punjab. Sangrur, Firozpur and Bathinda districts from Malwa region contributed more than ten per cent to the total fire counts, whereas, Majha and Doaba region districts contributed the least which is less than five per cent to the total fire counts of stubble burning in Punjab. This may be because of hilly terrain and diverse cropping system. Gurdaspur, Amritsar and Tarn Taran of Majha region are traditional Basmati growing regions. Most of the harvesting is done till late November with manual cutting which leaves low amount of residue in the fields and hence resulted in less burning. Patiala, Moga, Mansa and Ludhiana and Muktsar of Malwa region also contributed to five-ten per cent of total fire counts of stubble burning.

Another study conducted by Chhabra *et al.* (2019) summed the daily residue burning events in the districts of Punjab, Haryana and Uttar Pradesh over monitoring period (1st October to 30th November) for 2018. Sangrur district of Punjab had recorded the maximum number of stubble burning events. About 2,700 average burning events per district were observed in Punjab, while the figure was 440 and 90 in the Haryana and Uttar Pradesh respectively. Out of 22 districts, twelve districts namely; Barnala, Bhatinda, Faridkot, Fazilka, Firozpur, Ludhiana, Mansa, Moga, Muktsar, Patiala andSangrur from Malwa region and Tarn Taran from Majha region recorded higher stubble burning events than the average burning events per district (Figure 4). This clearly shows that Malwa region of Punjab contributes maximum to stubble burning in the state. Based on the above fact, it was selected as locale while **the three districts of the region; *viz.* Bathinda, Ludhiana and Sangrur were selected randomly for the study.**

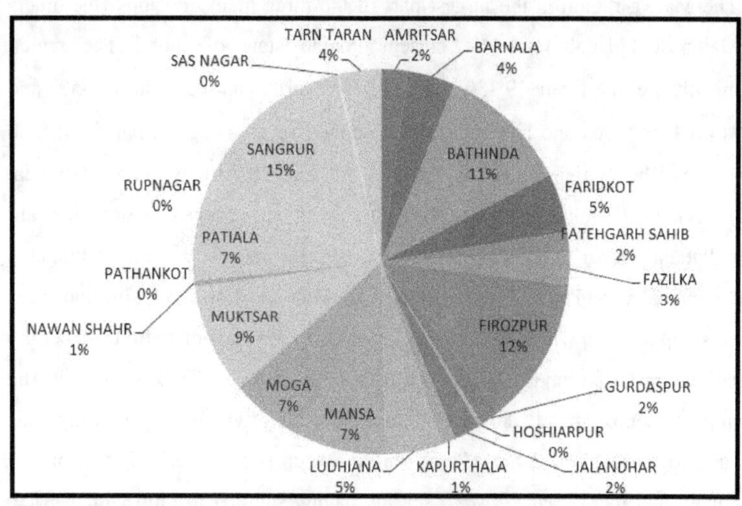

**Figure 3: District wise fire counts percentage in Punjab in 2018
(Kumar *et al.*, 2019)**

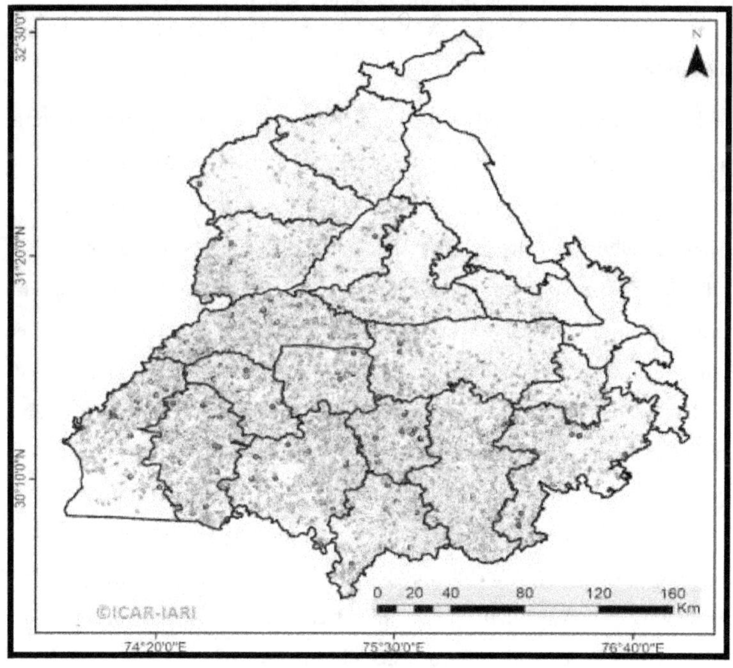

**Figure 4: Spatial distribution of active fire locations in Punjab in 2018
(Chhabra *et al.*, 2019)**

The Malwa region is the largest part of the three main divisions (the others being Majha and Doaba) of the present Punjab state of India. The region geographically lies between 29°-30' and 31°-10' North latitudes and 73°-50' and 76°-50' East longitudes and has been best described as the southern part of Punjab partitioned by the Sutlej River. It includes the districts of Barnala, Bathinda, Fatehgarh Sahib, Faridkot, Fazilka, Firozpur, Ludhiana, Mansa, Moga, Mohali, Muktsar, Patiala, Ropar and Sangrurthat comprises around 65 per cent of Punjab's total geographical area. The total population of the Malwa region is 14.3 million (52% of Punjab's population), with nearly 62 per cent being the rural population. As per the Statistical Abstract of Punjab (2011), out of a total area of 30200 sq. km. in the Malwa region, approximately 86.5 per cent is under cultivation. Climatically, this region has two seasons: the *Rabi (Harhi)* season, which is characterized by winter to mild summer, and the *Kharif (Sawani)* season, having summer to mild winter. Rice, wheat and cotton are the major crops grown in the region. A brief account of the selected districts has been given in the following sub-sections.

3.1.2.1 Bathinda

Bathinda district is located in the heart of the Malwa region, in the Central Southern part of Punjab. It is situated between 29°-45' and 30°-45'North latitude and 74°-30' and 75°-30' East longitude. Bathinda forms a part of Faridkot Revenue Commissioner's Division. The district shares its boundaries with Muktsar and Faridkot districts in the North and the West respectively, Mansa district in the South and Sangrur district in the East. According to 2011 Census, Bathinda covers an area of 3,353 sq. km and has a population of 13,88,525 which comprised of 7,43,197 males and 6,45,328 females. Thus, Bathinda district constitutes 17.3 per cent of Punjab's population and 6.7 per cent of its area. Among the districts, it ranks 15th in area and 13th in terms of population in the state. The average population size of village in the district is 3,187 which is greater than that of the state's average village population of 1,425. A total of 44 villages in the district are large-sized with a population of 5,000 or more. Bathinda district is relatively less urbanized as 36 per cent of its population resides in the urban areas while the corresponding figures of the state are 37.5 per cent. The sex ratio in the district is 868 which is lower than the state's figure of 895.

The district is further sub-divided into eight community development blocks; namely, Bathinda, Nathana, Talwandi Sabo, Sangat, Phul, Rampura, Maur and Bhagta Bhaika.

Over all, the climate in Bathinda district is dry. The district is mainly dependent on agriculture as 64.04 per cent of its population is reported to be residing in the rural areas during 2011 Census. Further, agriculture alone engaged 59.6 per cent of the district's main workers as per 2011 Census (cultivators 31.26 per cent, agricultural labourers 28.36 per cent). Thus, agriculture provides the single largest source of employment in the district though decline is noticeable during the last few decades. The *Kharif* and *Rabi* are two main harvesting seasons. The majorcrops of *Kharif* season include cotton, paddy, jawar, bajra, sugarcane, and groundnut. The principle *Rabi* crops grown in the district include wheat, gram, barley, some oil seeds, pulses, etc. Bathinda is also considered as cotton-growing belt of Punjab. Like other districts in Punjab, Bathinda district is also rapidly adopting rice-wheat cropping system along with the prevalent cotton-wheat cropping system. Rice crop is grown more in northern part of the district, where tube wells are the major sources of irrigation, whereas the cotton crop is grown mostly in central and southern parts of the district where the irrigation is widely provided by canals.

Village industries like handloom weaving, oil extraction by wooden *Kohlus*, manufacturing of agricultural implements, *jutti* making, *baan* making, jiggery *(gur)* and *shakkar* manufacturing, calico printing and *phulkari* making have been vogue in the rural areas. *Durries* (floor carpets) in floral designs are manufactured in the urban centres like Bathinda, Rampura Phul etc. There exists a colony of potters (*Kumhars*) at Bathinda which produce quality earthenwares, especially *surahis*. Day by day, industries are also developing in the backward areas of the district. The following kinds of industrial units in the small scale sector are now active in the district: Agricultural implements, Guru Gobind Singh oil refinery, sewing machines parts, steel rolling mill, wood and screws, electric goods, radio transistors and sound equipments, conduit pipes and plastic goods. A Regional Research Station of Punjab Agricultural University is also present in the district. Its thrust is to conduct research on cotton, wheat, pulses, vegetable and fruits (guava, grapes, orange, ber etc.) and also the seed production.

3.1.2.2. Ludhiana

Ludhiana is the most centrally located district, which falls in the Malwa region of the State of Punjab. The topography of the district is a typical representative of an alluvial plain and it owes its origin to the aggravational work of the Satluj. For administrative purpose it has been placed in the Patiala Division. It lies between 30°-34' and 31°-01' North latitude and 75°-18' and 76°-20' East longitude. It is bounded on the north by the river Satluj, which separates it from Jalandhar and Shahid Bhagat Singh Nagar districts. The river also forms its northern boundary with Hoshiarpur district. On other sides, it shares common boundaries with Rupnagar district in the East and Moga district in the West; and Sangrur and Fategarh Sahib in the South.

Ludhiana ranks third in terms of area and first in terms of population in the Punjab. The average population size of a village in the district is 1,615 is higher than that of the state which is 1,425. There are 32 large-size villages, having a population of 5,000 or above in the district. The district is highly urbanised as 59.2 per cent of its population resides in urban areas while that of Punjab is 37.5 per cent. The sex ratio in the district is 873 which is much lower than that of the state figures of 895. There are twelve community development blocks in the district, namely, Doraha, Machhiwara, Samrala, Khanna, Dehlon, Ludhiana-I, Ludhiana-II, Pakhowal, Sidhwan Bet, Jagraon, Sudhar, and Raikot.

Ludhiana is the most advanced agricultural district and plays an important role in directing the shape of agriculture in Punjab. It was among the foremost districts of the country where Intensive Agrcultural Development Programme was launched. According to the 2011 Census, agriculture is the single largest source of employment and livelihood in the district as it employed 36.7 per cent of total workers (main and marginal). The total geopraphical area of the district is 3767 sq. km. The district has three lakh hectares of net area sown, out of which almost 100 per cent is double cropped, and some area is even put to three crops a year. All of the area is irrigated, mostly by tubewells. The water table is going down at an alarming rate in Ludhiana district and even more seriously around the Ludhiana city blocks. There are 72,000 operational holdings, out of which, one-third belong to the small and marginal ones of less than two hectares.

There are two principal crop seasons in district namely *Kharif* and *Rabi*. The crops grown during *Kharif* season in the district are: paddy, maize, groundnut, sugarcane, cotton, pulses and chillies, etc. On the other hand, wheat, gram, barley, potatoes and oilseeds are the crops grown in *Rabi* season in the district. The district has the distinction of having the agricultural universities of the state; Punjab Agricultural University and Guru Angad Dev Veterinary and Animal Sciences University.

Role of Punjab Agricultural University

Punjab Agricultural University was established in 1962 to serve the state of erstwhile Punjab. It is modelled on the pattern of land grant colleges in United States. It has played a key role in increasing food production in Punjab state and ushering in an era of Green Revolution in India. The university has made notable contributions in increasing livestock and poultry production in the state. Over the years, Punjab Agricultural University have come forward to address the challenges faced by the farmers and holds several accolades in the field of agricultural technology generation and dissemination that proved to be of utmost importance to the farming community. The problem of stubble burning has also been the focal point of the research and extension in the University from last few years. With tremendous efforts, it has developed a machine called Modified Happy Seeder and a Super SMS attachment with harvest combine for paddy stubble management. The use of these machines facilitates the farmers to sow the successive crop without burning the paddy stubble. These machines have been made available to farmers at subsidized rates under various schemes. The messages of harmful effects and the alternative options of stubble burning have been propogated among the farming communities through various media by the university. These media include texts, videos, folk songs, drama, skits, plays, etc. The school children have also been included in the mass awareness campaigns and they propogate the messages by rallies, slogans and similar activities. The extension division of the university has been pro-active in sending these messages from time to time in different forms and through various events.

3.1.2.3. Sangrur

The district Sangrur is located in the Malwa track in the southern part of the state, between 29°-44' and 30°-42' North latitude and 75°-18' and 76°-13' East longitude. It is surrounded by the Ludhiana district in the north, Patiala in the east, Bathinda in the west, Faridkot in the northwest and Haryana state from the south. As per Census 2011, Sangrur district covers an area of 3,625.0 sq.km and has a population of 16, 55,169 which comprised of 8, 78,029 males and 7, 77,140 females. It accounts for 7.2 per cent of Punjab's geographical area and 6.0 per cent of its population. Among the districts in the State, Sangrur ranks second in area and fourth in population. The average population size of village in the district was 2020, which is greater than that of the state figures of 1425. More than 25 villages in the district are large sized with a population of 5000 or more. Sangrur district is relatively less urbanized than the state as its 31.2 per cent population is in urban areas. The sex ratio in the district is 885, which is also lower than that of the state figures of 895. The district Sangrurhas nine C.D. blocks namely Malerkotla-I, Malerkotla-II, Sherpur, Dhuri, Bhawanigarh, Sangrur, Sunam, Lehara Gaga and Andana.

According to Census, 2011, 68.8 per cent of population of Sangrur district resides in the rural area and 31.2 per cent in urban area. However, there has been marked shift in the economic composition of its population. Less than one-half (48.1%) of its population is engaged in agriculture (28.8% as cultivators and 19.3% as agricultural labourers). Similar to other districts, *Khaif* and *Rabi* are the two main crop seasons. The major *Kharif* crops include paddy, jawar, bajra, sugarcane and groundnut. On the other hand, *Rabi* crops include wheat, barley, gram, oil seeds, etc.This district is marked with relatively low rainfall and thus, irrigation has a vital role. Tubewells and wells are the main source of irrigation in the district. Well irrigation is a very old agricultural practice in the district where bullocks and camels provide the main source of power for running the Persian wheel. However, the introduction of pumping sets run by diesel and hydro-electric power has considerably increased in the district.

Malerkotla, Dhuri, Sangrur, Sunam and Moonak are the main centres of trade in the district. Agricultural commodities constitute the main items of export. They include wheat, paddy, gram, cotton, jaggery (*gur*), sugar, etc. Among the non-agricultural items, export of bicycle, sewing machine parts, leather goods, etc. finds mention in exported commodities.

The Table 3 describes the selected districts of Malwa region at a glance from various aspects:

Table 3: Bathinda, Ludhiana and Sangrur districts at a glance

S. No.	ITEMS	UNIT	STATISTICS			
			Punjab	Bathinda	Ludhiana	Sangrur
A	**Area**					
1.	Total	km^2	50362	3353	3578	3625
B	**No. of Blocks**					
1.	Total	No.	150	9	13	10
C	**No. of villages**					
1.	Total	No.	12,581	281	907	571
(i)	Inhabited	No.	12,168	279	885	564
(ii)	Uninhabited	No.	413	2	22	7
D	**No. of households**					
1.	Normal	No.	54,86,851	2,72,530	7,13,493	3,17,391
2.	Institutional	No.	16,367	905	2,198	879
3.	Houseless	No.	9,853	467	1,135	600
E	**Population**					
1.	Total Population	No.	2,77,43,338	13,88,525	34,98,739	16,55,169
(i)	Males	No.	1,46,39,465	7,43,197	18,67,816	8,78,029
(ii)	Females	No.	1,31,03,873	6,45,328	16,30,923	7,77,140
2.	**Urban Population**	No.	1,03,99,146	4,99,217	20,69,708	5,15,965
(i)	Males	No.	55,45,989	2,68,713	11,14,372	2,73,376
(ii)	Females	No.	48,53,157	2,30,504	9,55,336	2,42,589
3.	**Rural Population**	No.	1,73,44,192	8,89,308	14,29,031	11,39,204
(i)	Males	No.	90,93,476	4,74,484	7,53,444	6,04,653
(ii)	Females	No.	82,50,716	4,14,824	6,75,587	5,34,551
F	**Sex Ratio**					
1.	Total	Females per thousand males	895	868	873	885

			Punjab	Bhatinda	Ludhiana	Sangrur
(i)	Rural	-do-	907	874	897	884
(ii)	Urban	-do-	875	858	857	887
G	**Literacy Rate**					
1.	Total	%	75.84	68.28	82.20	67.99
(i)	Males	%	80.44	73.79	85.98	73.18
(ii)	Females	%	70.73	61.94	77.88	62.17
H	**Categorization of Workers**					
1.	Cultivators	No.	19,34,511	1,41,535	1,40,737	1,57,191
(i)	Males	No.	17,53,359	1,26,449	1,28,748	1,47,916
(ii)	Females	No.	1,81,152	15,086	11,989	9,275
2.	Agricultural Labourers	No.	15,88,455	1,28,382	98,646	91,799
(i)	Males	No.	12,39,445	95,085	81,595	78,165
(ii)	Females	No.	3,49,010	33,297	17,051	13,634
3.	Workers in household Industry	No.	3,85,960	14,938	71,667	18,578
(i)	Males	No.	2,49,294	8,812	49,629	13,240
(ii)	Females	No.	1,36,666	6,126	22,038	5,338
I	**Area, Production and Productivity of major crops (2017-18)**					
	Crop	**Unit**	**Punjab**	**Bhatinda**	**Ludhiana**	**Sangrur**
1.	**Wheat**					
(i)	Area	'000 ha	3,512	254	252	288
(ii)	Production	'000 MT	17,830	1,345	1,296	1,599
(iii)	Productivity	kg/ha	5,077	5,295	5,144	5,552
2.	**Rice**					
(i)	Area	'000 ha	3,103	167	259	287
(ii)	Production	'000 MT	19,136	1,137	1,721	2,019
(iii)	Productivity	kg/ha	6,167	6,806	6,646	7,036

Source: District Census Handbooks of Bathinda, Ludhiana and Sangrur, 2011

3.2. SAMPLE AND SAMPLING PROCEDURE

Sampling is the statistical process of selecting a part of population, also known as a "sample" for the purpose of making observations and statistical inferences about that population. The sampling procedure adopted for the present studyis described below:

3.2.1. Selection of Blocks: One block from each selected district, thus, a total of three blocks were selected through Simple Random Sampling without replacement. Out of nine blocks of Bathinda district, Bathinda block was selected; while out of thirteen blocks of Ludhiana district, Khanna block was selected by Simple random sampling method. Similarly, out of ten blocks of Sangrur district, Sunam block was selected randomly. Thus, a total of three blocks from three districts of Malwa region were selected for the study. A brief description of the selected blocks has been presented in the Table 4.

Table 4: Description of the blocks selected for the study

S.No.	ITEMS	UNIT	STATISTICS		
			Bathinda	**Khanna**	**Sunam**
A	No. of villages	No.	29	73	41
B	Current Population	No.	90,615	94,160	1,38,504
C	No. of families	No.	17,972	18,717	26,239
D	Net Sown Area	Hectares	23,687	16,964	36,550
E	Gross sown Area	Hectares	44,419	32,162	73,016
F	Area Under Cultivation				
(i)	Wheat	Hectares	20,301	13,689	34,326
(ii)	Rice (paddy)	Hectares	11,476	14,296	31,888
(iii)	Cotton	Hectares	9,435	0	1,116
(iv)	Sugarcane	Hectares	4	451	85
(v)	Maize	Hectares	7	27	51
(vi)	Oilseeds	Hectares	229	71	174
(vii)	Vegetables	Hectares	357	1,803	211
(viii)	Fodder	Hectares	1,981	1,619	4,918

Source: District Level Reports of Bathinda, Ludhiana and Sangrur, 2017-18

The data in Table 4 shows that the selected blocks are the major growers of rice and wheat. This also indicates that the farmers in the selected blocks follow rice-wheat cropping pattern. The selection thus meets the the basic requirement of the study.

3.2.2. Selection of Villages: After the selection of blocks, a list of villages was prepared for each block. The total number of villages across the selected blocks was found to be 143. Out of which, 29 villages belonged to Bathinda block of Bathinda district, 73 villages to Khanna block from Ludhiana district and rest 41 villages were from Sunam block of Sangrur district. These villages constituted the sampling frame for the study. Further, two villages from each block were selected through Simple Random Sampling without replacement. As a result, villages named Khialiwala and Amargarh from Bathinda block; Goh and Lalheri from Khanna block; and Lakhmirwala and Kharial villages from Sunam block were selected randomly for the study. To verify that the major proportion of population in the selected villages were engaged in agriculture, preliminary field visits were also conducted by the researcher under the guidance of local Agricultural Officers.

3.2.3. Selection of Respondents: The basic criteria for selection of farmers as the respondents for the study was that the farmer must have been following the rice-wheat cropping pattern for atleast past five years.

There are different formulae for determination of appropriate sample size when different sampling techniques are used. For selecting farmers from the villages, appropriate sample size was estimated by using Cochran's formula, which is given below:

$$n_0 = \frac{Z^2 pq}{e^2}$$

Where,

$n_0 =$ sample size

$Z=$ selected critical value of desired confidence level

$p=$ estimated proportion of an attribute that is present in the population and $q= (1-p)$

$e=$ desired level of precision

Assuming the variability to 20 per cent (i.e. q=0.2) and taking 95% confidence level with ±5 per cent precision, the required sample size was calculated as follows:

As, q = 0.2

p =1-0.2 = 0.8

e = 0.05

z =1.96

$$n_0 = \frac{(1.96)^2(0.8)(0.2)}{(0.05)^2} = 245.86 = 246(approx.)$$

Therefore, 246 farmers were selected as respondents through proportional allocation for the fulfilment of pre-determined objectives. Table 5 describes the selection of the respondents according to the farmer population in the villages.

Table 5: Selection of the respondents

S. No.	Blocks	Selected Villages	Total Population	Farmers Population	Respondents for study
1.	Bathinda	Khialiwala	3091	447	67
		Amargarh	1151	235	36
2.	Khanna	Goh	2879	268	40
		Lalheri	2278	141	21
3.	Sunam	Lakhmirwala	1719	171	26
		Kharial	2882	370	56
	TOTAL		14000	1632	246

The data in Table 5 shows that the maximum number of respondents was selected from Khialiwala village (67) of Bathinda block, followed by Kharial village (56) of Sunam block and Goh village (40) of Khanna block. The number of respondents selected from village Amargarh from Bathinda block were 36 while, 26 and 21 respondents were selected from villages Lakmirwala of Sunam block and Lalheri from Khanna block respectively.

The overall sampling plan followed for the study is depicted in Figure 5.

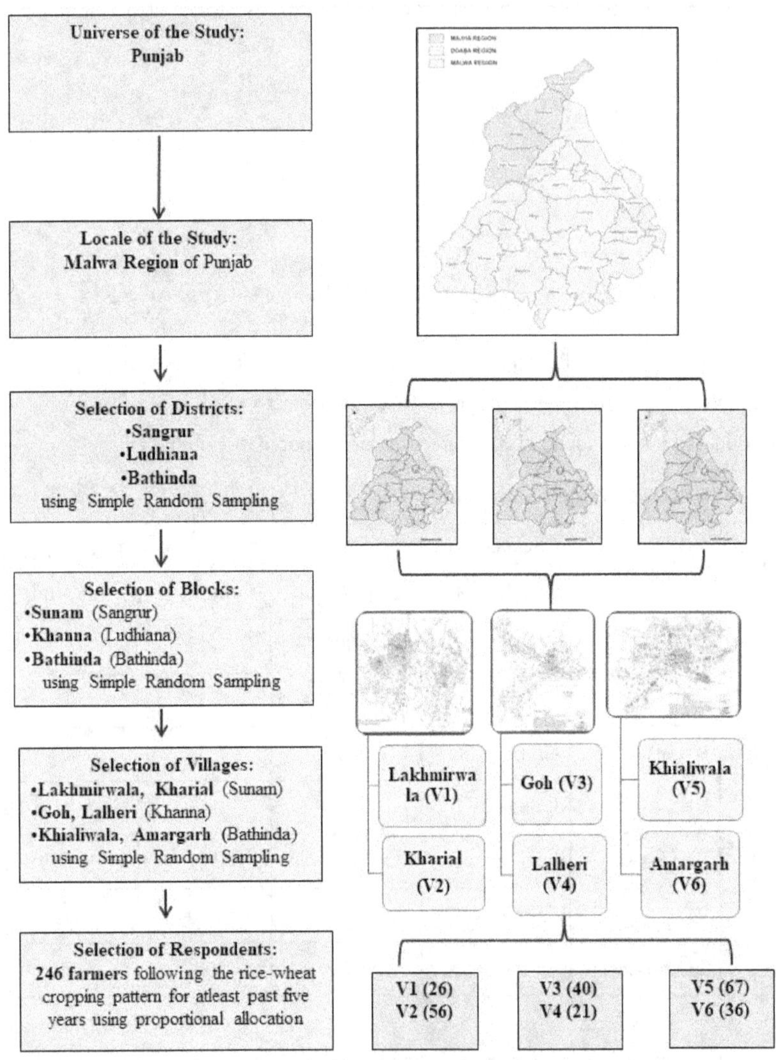

Figure 5: Sampling procedure

3.3 DESCRIPTION OF THE STUDY VILLAGES

The study was conducted in six villages namely Khialiwal, Amargarh, Goh, Lalheri, Lakhmirwala and Kharial located in Malwa region of Punjab. The general statistics of the villages have been given in Table 6. The data in Table 6 indicates that out of the selected villages, Khialiwal from Bathinda block was the largest village in terms of area with 1045 hectares while Lalheri village from Khanna block was the smallest in terms of area which had 409 hectares. Khialiwal village from Bathinda block had 553 households which was the highest among the selected villages, while Amargarh village from Bathinda block had the least number of households, 220.

Table 6: General Statistics of the selected villages

S. No.	Item	Bathinda		Khanna		Sunam	
		Khialiwal	Amargarh	Goh	Lalheri	Lakhmirwala	Kharial
1	**Area (hectares)**	1,045	496	606	409	535.79	794
2	**No. of households**	553	220	548	437	324	550
3	**Population**	3,091	1,151	2,879	2,278	1,719	2,882
4	**Literacy rate (%)**	55.32	59.60	73.25	71.68	59.51	54.92
5	**No. of cultivators**	447	235	268	141	171	370

Village Khialiwal was the most populous village having 3091 residents while Amargarh village also from Bathinda block had the least number of residents 1151 residents. It was also noted that Goh village from Khanna block had 73.25 per cent literacy rate and Kharial village had 54.92 per cent literacy rate which were recorded to be the highest and the lowest rates respectively among the selected villages. The maximum number of cultivators (447) were in Khialiwal village while Lalheri from Khanna block had the least number of cultivators, i.e. 141.

The consolidated description of the villages under study from different contexts has been given below:

3.3.1. People

As the study was concentrated in Malwa region of Punjab, all the sampled villages under study had more or less similar strata of people living in them. Almost all the selected villages were dominated by *Jatt Sikhs* with varying proportions of *Khatris, Banias* and *Mazhabi Sikhs* scattered widely in the villages. The *Jatt Sikhs* are commonly called as '*sardarji*' while some communities refer to them as '*zimidars*' meaning 'land lords'. They usually own large landholdings and farming is their primary occupation. The *Khatris* are said to be originated as *Kshatriyas*, meaning 'warriors', referring to the second highest class of Hindu *Varna* system. But with time, they have become an influential community in the field of trade, commerce and agriculture. *Banias* are the traders, classified as '*Vaishyas*' according to the Hindu Varna system. They are commonly called as '*Lala*', '*Seth*'or '*Sahukar*' by the members of other communities. They are rich, held various kinds of businesses and lent money to others. *Mazhabi Sikhs* are considered as the members of lower caste. They follow Sikhism and agriculture as their major occupation. Residing in the Malwa region, people had more or less similar dialect of Punjabi language. The form of Punjabi language used in general speaking in this regon is called as *Malwai* which is different from *Majhi* or *Majhel* and *Doabi*, the dialects of other major regions of Punjab. They all are written in Gurumukhi script but the usage of some words, accent and pronunciation are different.

3.3.2. Settlement pattern and Housing

The lifestyle of people or communities is reflected by the environment they create around themselves. The type of housing conditions in which they live in is one such indicator. The settlement pattern in the selected villages was basically of two kinds: the big farm houses, in which the houses were located within the agricultural fields and the typical medium-sized houses made side by side, facing each other with narrow *gullies* passing through them. All of the houses were *pucca* in nature, however, the nature of investment on the beautification of houses varied across the villages and the classes of people residing in them. All the houses were electrified. However, one common thing was observed that almost all the houses had a courtyard called '*vedha*' which was kept open and used for various purposes. Majority of people had kept milch animals usually buffaloes (*majjh*) and cows (*gaa*). The cattle were usually kept in a separate shed made in one corner of these courtyards. The milk or *dudh* obtained from the cattle was consumed within the families and various dairy products like *ghee, malai, paneer, lassi, makkhan,* etc. were made for daily home consumption. The colustrum or commonly called as '*boli*' delivered by the newly milking cow, after feeding the calves, was consumed within the families. It is considered highly nutritious especially for children. In some villages, it was noted that excessive milk was sold to the people who did not own cattle and in nearby villages; while some villages had a tradition of giving the milk out to any needy who comes to the house, free of cost. The cow dung was used to make dung cakes or *patthiyan* which were used as fuel in traditional *tandoors* which was also a common feature in the houses. The excessive cow dung was also used in the fields as rotten manures. Some households also had pegions, which they trained for participating in the local competitions organized in the villages. Some had kept dogs and horses as their pets. Many households had kitchen gardens which are basically utilized for home consumption.

All the villages had atleast one *Gurudwara*, the common religious place in the village where people went to offer prayers or commonly known as '*ardaas*' on a daily basis. The day begins quite early in the villages with the sound of *paathi*, the preacher or reader in the Gurudwaras reciting the holy sermons and hymns. This time is called '*amrit vela*' and it usually clocks around 4 am, early in the morning. Usually, the elders in the village wake up at this time and perform their daily chores and utilize their time in '*nit-niyam*', the daily holy rituals either in Gurudwaras or at their homes itself. After that, the males usually go to the fields and their respective shops or businesses, children to their schools and females perform their various household responsibilities including taking care of domestic animals. Many households had joint families while there were many others who were partitioned and were living separately. As told by many elders of the villages during interaction, the villages are now experiencing a transition phase where new generation kids are looking for better livelihood options other than farming. It was also noticed that new generation parents were reluctant to keep their children in the farming occupation because over time it has proved to be less profitable and even a matter of loss to some, since past few years. Not only this, in the last four-five years, many parents from the selected villages have sent their younger generation to other countries specifically to Australia, New Zealand and Canada for better educational and employment opportunities.

3.3.3. Economy

Access to livelihood is the basic requirement of every human being. The people in the selected villages for the study were mostly dependent on agriculture. As the villages were well-electrified, tube wells were readily available which serves as the major source of irrigation. Major crops grown in the selected villages were wheat, paddy, cotton, maize, mustard, etc. Overall, the area under wheat and paddy crops dominated in the villages of Khanna and Sunam blocks, while cotton had a significant area in comparison to wheat and paddy in the villages of Bathinda block. The crops were grown mostly for commercial purpose. After harvesting, the produce was sold to local middlemen (*adatiyas)* or grain merchants in the nearby *mandis*. All the selected villages were in five-twelve km radius of the nearby towns of Bathinda, Khanna and Sunam where they usually sold their produce.

It was observed that overall; a wide range of modern and advanced farm implements could be seen in the villages, especially in the villages of Khanna and Sunam blocks. Implements and machines like tractors, Mould-bold plough, land levellers, rotavators, mulchers, choppers, etc. were available in the villages. The implements were privately owned by some big individual farmers and also owned in groups by some small farmers. Others who could not afford such costly implements manages their farm by using these implements on rent and lease basis. For managing the paddy stubble after harvesting, Happy Seeders and Super Seeders were used by the farmers machines specifically manufactured for the purpose of avoiding stubble burning. It was arguable that although the usage of machines and implements added to the overall production cost, but it also saved much of the time of the farmers.

The people in the selected villages of Bathinda block had comparatively more diverse set up of economic activities which included other livelihood options than field cultivation. Apart from agriculture, various other prevalent activities in these villages included brick-buliding, commonly known as '*eent di bhatti*' or just '*bhatta*'; poultry farming, small confectionaries, *kirana* shops, dairies, etc. Apart from these, many members of communities like *Khatris, Bania* etc. in the selected villages owns and runs their businesses and shops in the nearby towns. Their businesses includes garment shops, mobile phone shops, implements repairing worskshops, grain merchants, tractor and other auto-dealerships, fruit and vegetable merchants, jewellery shops, etc. These businesses served as primary occupations to many but secondary occupations to some people who were actively involved in farming. Many people from the villages who decided to change their business or have migrated abroad or to other cities, had given their land on lease to other farmers. The well settled children of families who flew away to abroad from the villages sent money to their families. Some girls and women run beauty parlours exclusively for women in the villages. Others gave tuitions to school going children from around the village while many relied on sewing, knitting and stitching skills for additional family income.

3.3.4. Religion and Festivals

Punjab is the land of origin of the world's ninth largest followed religion called as Sikhism. The adherents of Sikhism are known as *Sikhs*, meaning 'students' or 'disciples' of the Guru. As mentioned earlier, all the selected villages are dominated by the Sikhs. Sikhism is based on the spiritual teachings of Guru Nanak Dev Ji, the first Guruand the nine Sikh Gurus that succeeded him. The tenth Guru, Guru Gobind Singh Ji named the Sikh scripture Shri *Guru Granth Sahib* as his successor, terminating the line of human Gurus and establishing the scripture as the eternal, religious spiritual guide for Sikhs. The Sikh scripture opens with the *Mul Mantar*, fundamental prayer about *ik onkar* (One God). The core beliefs of Sikhism, articulated in the *Guru Granth Sahib*, include faith and meditation on the name of the one creator; divine unity and equality of all humankind; engaging in *seva* (selfless service); striving for justice for the benefit and prosperity of all; and honest conduct and livelihood while living a householder's life. It teaches followers to discard the "Five Thieves" (i.e. lust, rage, greed, attachment, and ego) that are present in them. Most of the elders of the study villages are strong believers of the religion and follow the teachings whole heartedly. However, some of them believed that the younger generations are getting influenced by westernization and not adhering to the holy teachings as their forefathers have.

Punjab is the land where festivals and fairs/Melas are synonymous with joy and gaiety. *Baisakhi* is one of the most important festivals of Sikhs. Sikhs celebrate it because on this day, the tenth Guru, Guru Gobind Singh, inaugurated the Khalsa, the 11th body of Guru Granth Sahib and leader of Sikhs till eternity. It includes the *Nagar Kirtan*, the processional singing of holy hymns throughout community. Traditionally, the procession is led by the saffron-robed *Panj Pyare* (the five beloved of the Guru), who are followed by the Guru Granth Sahib, the holy Sikh scripture, which is placed on a float. *Bandi Chor Diwas* has been another important Sikh festival in its history. Sikhs celebrate Guru Hargobind's release from the Gwalior Fort, with several innocent kings who were also imprisoned by Mughal Emperor *Jahangir* in 1619. This day continues to be commemorated on the same day of Hindu festival of Diwali, with lights, fireworks and festivities. *Gurpurabs* are celebrations or commemorations based on the lives of the Sikh Gurus. They tend to be either birthdays or celebrations of Sikh martyrdom. All ten Gurus have *Gurpurabs* on the *Nanakshahi* calendar, but it is Guru Nanak and Guru Gobind Singh who have a *Gurpurab* that is widely celebrated in Gurudwaras and homes. The martyrdoms are also known as a *Shaheedi Gurpurabs*, which mark the martyrdom anniversary of Guru Arjan Dev and Guru Tegh Bahadur.

Lohri is a popular Punjabi festival, which commemorates the passing of the winter solstice as Lohri was originally celebrated on winter solstice day, being the shortest day and the longest night of the year. *Mela Maghi*, occursin the month of Magh according to *Nanakshahi* calendar is one of the most important fairs and the most important of all religiously significant gatherings of the Sikhs. These festivals and fairs are celebrated throughout Punjab and everywhere where the Sikhs reside. The members of other communities like *Khatris* and *Banias* also take part in these festivals and fairs with a lot of enthusiasm, which shows the solidarity of people living together in harmony and peace.

3.3.5. Connectivity and transportation

Connectivity with the developed or prosperous regions lays a great impact on the progress of a village; be it in terms of information, services or transportation. The roads of the selected villages were mostly well built, especially in the villages of Bathinda and Sunam blocks. However, the roads inside the Goh village of Khanna block were yet to be constructed. Still, all the selected villages were well connected with the nearby towns. People had range of vehicles, from bicycles to motor cycles, scooties, cars, tractors, trollies, etc. under their possession. Motor cycles were the most common source of transport, which was used by the people of all age-groups for various purposes. Youngsters preferred Royal Enfield Bullet which was a common sight in the households. Some young *Jatt Sikhs* in village Kharial even had branded cars like Audi, Mercedes and BMW. On the other hand, some of the elders used to travel on foot or by bicycle to the fields and inside the villages. Girls and women of the village used scooties for commuting in and around the villages. The young farmers in the villages had an interesting craze for putting high-volume speakers on their tractors and playing old classic as well as latest Punjabi songs while travelling and even working on the fields. They had large-sized trollies modified with alloy-wheels. A frequent bus service was also present from town to various villages in Bathinda and Khanna blocks. Overall, all the villages were well connected to the nearby towns in terms of transportation.

3.3.6. Leisure and Recreation

Leisure and recreational activities hold an important role in the shaping of the society as a whole. Most of the elders of the village pay regular visits to the Gurudwaras as they found peace in the soulful environment. Some of them performed *seva* as community service which included various chores like cleaning, etc. The occasions of child birth, marriages, etc. in the villages also featured various *akhand paaths* for peace and happiness. These were usually followed by the *langars*, the open community food service to the members of the villages. Various merry occasions included *shabeel,* the sweetened water distribution among the commuters alongside the roads of the village. The people in the villages were extremely fond of *Kushti* or wrestling. Various competitions were organized at different levels and the winners were handsomely rewarded. The youngsters of the villages spent their time playing volleyball, cricket and *kabaddi* in the evenings. Hockey was also preferred by some boys. Some people trained pigeons for intra and inter-village level competitions. The competition is commonly known as *kabootar baazi*, which took place usually as a leisure activity but was also an interesting feature of fairs and festivals which drew large masses from various villages and even across the districts. The pigeons are made to fight each other according to a set of rules. The winner gets a reward in the end. The women in the villages usually spent their leisure time watching daily soaps on

TV. Most of the teenagers used social media applications like WhatsApp, Facebook, YouTube, SnapChat, Hike, etc. for chatting and video calling. Some young boys and girls utilized these platforms for learning purposes as well. A considerable amount of people, from all the age-groups and genders were found making funny videos on social media platforms like TikTok. They invest a significant amount of time in making these videos and circulate them among family and friends. In many parts, an alarming trend in the substance abuse of *afeem* and *bhukki* as recreational activity was also noted among the youngsters, which often turns into addiction.

3.3.7. Organizations in the villages

The selected villages had few groups, committees, societies and organizations at their level which looked upon various kinds of activities in their respective villages. The first of the kind were the *Self-Help Groups* formed by women of two selected villages in Bathinda block and one village in Sunam block of Sangrur district. The three groups were registered under the names as Guru Nanak Dev Self Helf Group, Guru Gobind Singh Self Helf Group and Baba Jivan Singh Self Helf Group. They were formed to empower the women of the villages and were functional from past three-five years. Another type of village organization which was found in all the selected villages was the *Gurudwara committee.* It looked after the proper functioning and maintenance of the Gurudwaras. These committees were also responsible for the celebration of various festivals and fairs in the respective villages. The third kind of committee was *Sports clubs* which were responsible for the recreational activities in the villages such as organization of sports activities like volleyball, kabaddi, *kushti* tournaments. Sports clubs were common in the selected villages of Khanna block. There was also a group of farmers which were commonly called as societies. These *farmer societies* were officially registered and could avail the benefits of the subsidies in purchasing farm machineries as part of various government schemes. The primary objective of these societies was to offer tractors and farm implements to farmers on rent who owned small scaled tractors or could not afford at all. The farm machineries on which the subsidies were available included Happy Seeder, rotavator, mulcher, plough, etc. in various combinations. To check the practice of stubble burning in the state, the government provides these farm machineries and implements on 80 percent subsidy to these societies and on 50 per cent subsidy to the individual farmers. These farmer societies have a defined heirarchial structure which has one of the farmer representatives known as the *Sachiv* or Secretary at village level, the Area Registrar at block level and District Registrar at the district level. The fund flow was regulated by these officials.

3.4.   RESEARCH DESIGN

A research design is the "blueprint" of the systematic plan to study a scientific problem and which specifies how data should be collected and analysed. As defined by **Kerlinger (2004)**, research design is a "plan, structure and strategy of investigation so as to obtain answers to research questions and to control variance". This includes devising an appropriate sampling strategy, operationalizing constructs of interest and selecting a research method. In the present study, descriptive design was used which includes surveys and fact-finding enquiries of different kinds. The major purpose of descriptive research is description of the state of affairs as it exists at present and also shows cause and effect relation between non-manipulated variables.

3.5.   SELECTION OF VARIABLES AND THEIR MEASUREMENT

The focus of the study was to study the stubble burning behaviour of the farmers and factors responsible for the same. For this, the Theory of Planned Behaviour was used which has four variables, defined and operationalized later in the following section. Depending on their intended use, variables may widely be classified as independent or dependent variables. Variables that explain the occurrence or cause of other variables are called independent variables while those that are explained by or caused by other variables are dependent variables.

The variables under the present study were selected on the basis of extensive review of literature related to the topic of research after thorough consultation with experts. The variables which accounted to have most relevance to the present investigation were finally selected and included in the study. All the selected variables for the study have been listed in the Table 7:

Table 7: Selected Variables and their Measurement

	Independent Variables	Measurement
A	**Socio-personal, economic and communication variables**	
1	Age	Categories were made
2	Gender	Man/Woman
3	Education	Schedule was developed
4	Land Holding	Classification of Statistical Abstract of Punjab (2019)
5	Annual Income	Schedule was developed
6	Livestock possession	Schedule was developed
7	Type of Farming	Categories were made
8	Information seeking Behaviour	Bhairamkar (2009) with modifications
B	**Psychological and situational variables**	
1	Innovativeness	Jaussi and Dionne (2003) with modifications
2	Risk Orientation	Supe (2007) with modifications
3	Scientific Orientation	Supe (2007) with modifications
4	Ecological Consciousness	Roy (2015)
5	Economic motivation	Modified scale of Supe (2007)
6	Awareness about stubble management measures	Schedule was developed
7	Cropping Intensity	Formula was used
C	**Theory of Planned Behaviour Variables**	
1	Attitude towards Stubble Burning	Wang *et al.* (2018) with modifications
2	Subjective Norms associated with Stubble Burning	
3	Perceived Behavioural Control related to Stubble Burning	
4	Behavioural Intention regarding Stubble Burning	
5	Stubble Burning Behaviour	

3.6. OPERATIONALIZATION OF VARIABLES

Operationalization is the process of designing precise measures for abstract theoretical constructs or variables. An operational definition assigns meaning to a construct or a variable and describes how it will be measured according to the purpose of the study. The operational definitions for the selected variables are given as following:

3.6.1. Age

It referred to the chronological age of the respondent at the time of investigation, expressed in years rounded off. On the basis of this criterion, using mean (38.07) and standard deviation (12.14), the respondents were classified into three categories as young, middle-aged and old groups.

S. No.	Age	Range (years)
1	Young	<26
2	Middle-aged	26-50
3	Old	>50

3.6.2. Gender

It referred to the differences in man and woman in terms of role and status in society, values and attitude. It also referred to the cultural meanings attached to being masculine and feminine, which influences personal identities. The gender of the respondents was classified as male, female.

S. No.	Gender
1.	Male
2.	Female

3.6.3 Education

It referred to the level of formal school education of the respondents in terms of years he/she attended the school/college successfully. The respondents were categorized as illiterate, functionally literate, primary, high school, intermediate, graduate and post graduate.

S. No.	Education
1	Illiterate
2	Functionally literate
3	Primary education
4	High school
5	Intermediate
6	Graduate
8	Post graduate

3.6.4. Land holding

It referred to the total number of acres of land cultivated by an individual respondent. The respondents were classified into following five categories as mentioned in the Statistical Abstract of Punjab (2019):

S. No.	Size of Land holding	Range (acres)
1	Marginal	<2.5
2	Small	2.5-5
3	Semi-medium	5-10
4	Medium	10-25
5	Large	>25

3.6.5. Annual Income

It referred to the total earnings of the respondent in a year from cultivation of paddy and wheat without deducting the cost of cultivation or other miscellaneous costs incurred during the whole cultivation process. The respondents were classified into three categories of 'low', 'medium' and 'high' using minimum annual income (Rs. 50,000) and maximum annual income (Rs. 12,50,000).

S. No.	Annual Income	Range (Rupees)
1	Low	50,000 - 4,50,000
2	Medium	4,50,000 - 8,50,000
3	High	8,50,000 - 12,50,000

3.6.6. Livestock possession

It referred to the type of livestock possessed by the respondent. The types of livestock possessed by the respondents were categorized into poultry animals (chicken), smaller milch animals (Sheep or Goat), larger milch animals (Cow or buffalo) and draught animals (oxen).

S. No.	Livestock possession
1	Poultry animals
2	Smaller milch animals (Sheep or Goat)
3	Larger milch animals (Cow or buffalo)
4	Draught animals (Oxen)

3.6.7. Type of farming

It referred to the kind of farming followed by the respondent; i.e. specialized, diversified and mixed. Specialized farming referred to only one kind of farm business such as raising food crops. The motive is to earn maximized profits. Diversified farming is that kind of farming in which a farmer is engaged in a multitude of farm enterprises. Its purpose is to become self sufficient. Mixed farming is a special case of diversified farming which includes combining of two independent agricultural enterprises on the same farm in order to increase the income from different sources. The respondents were classified on the basis of frequency and percentage of the type of farming undertaken by them.

S. No.	Type of Farming
1	Specialized
2	Diversified
3	Mixed

3.6.8. Information Seeking Behaviour

It referred to the frequency of different sources *viz.* personal localite, personal cosmopolite, mass media and extension eduction methods used for obtaining information related to stubble burning and various agricultural operations. The extent of use of information sources was measured by taking into consideration all the possible sources available to the respondent. To measure this aspect scale developed by **Bhairamkar (2009)** was used and responses were collected on three-point continuum i.e. Regular (2), Occasional (1) and Never (0) respectively. The personal localite sources in the scale included friends, neighbours, relatives, progressive farmers and local leaders. Agriculture Officer (AO), Block Development Officer (BDO), Field Officer (Bank), KVK-Subject Matter Specialists (SMS) or Scientist, Input Retail Shops and Company Agents were the personal cosmopolite sources. The mass media sources included newspaper, farm magazines, television, radio, mobile phones and nataks/plays. Lastly, the extension education methods in the scale involved meetings, demonstrations, group discussions, field trials, kisan mela and workshops.

Separate analysis was done for all four categories of sources and based on weighted mean scores; the sources were ranked individually. In the end, taking into account the cumulative score of the respondents from all the sources, they were classified into following categories based on mean (48.87) and SD (3.38).

S. No.	Information seeking behaviour	Range (Score)
1.	Low	<46
2.	Medium	46-52
3.	High	>52

3.6.9. Innovativeness

It is the degree to which a respondent is relatively earlier in adopting new ideas in relation to agriculture and allied activities. The scale given by **Jaussi and Dionne (2003)** was taken into consideration and modified for the study with respect to stubble management measures. The scale consisted of statements regarding interest of the respondent in adoption, likeliness to adopt first and exposure about new information regarding stubble management measures. The scale consisted of five statements, of which four statements were positive while one was negative. The response were accorded on five point continuum ranging from strongly agree to strongly disagree. The scoring procedure used was as follows:

S.No.	Statements	Response				
		SA	A	UD	DA	SDA
1	Score for positive statement	5	4	3	2	1
2	Score for negative statement	1	2	3	4	5

SA=Strongly Agree, A=Agree, UD=Undecided, DA=Disagree, SDA= Strongly Disagree

After calculating their cumulative score, the respondnets were grouped into three categories namely 'low, 'medium' and 'high' by using mean (19.97) and SD (2.60).

S. No.	Innovativeness	Range (Score)
1.	Low	<17
2.	Medium	17-23
3.	High	>23

3.6.10. Risk Orientation

It referred to the degree to which a respondent is oriented towards risk and uncertainty for using new technology related to stubble management. The scale developed by **Supe (2007)** was modified for the study. The scale consisted of items which covered the aspects like willingness to take risks for environment, cost effectiveness, award of penalties for stubble burning, etc. The responses were recorded on five-point continuum ranging from Strongly Disagree (1) to Strongly Agree (5). The respondents were categorized into three categories namely 'low', 'medium' and 'high' based on mean (21.64) and SD (2.55).

S. No.	Risk Preference	Range (Score)
1.	Low	< 19
2.	Medium	19-25
3.	High	> 25

3.6.11. Scientific Orientation

It is the degree to which an individual is oriented in accordance with the procedures, methods, conduct and accepted conventions of modern science while adopting new agricultural techniques. The scale developed by **Supe (2007)** was modified for the study. The scale consisted of items addressing aspects like difference between performance or outcomes of traditional and scientific methods of agriculture, their affect on incomes and other aspects of life, etc. The responses were recorded on five-point continuum ranging from Strongly Agree (5) to Strongly Disagree (1). After calculating their cumulative score, the respondnets were grouped into three categories namely 'low, 'medium' and 'high' by using mean (22.48) and SD (2.73).

S. No.	Scientific Orientation	Range (Score)
1.	Low	<20
2.	Medium	20-25
3.	High	>25

3.6.12. Ecological consciousness

It is the degree to which an individual feel directly related and engaged with the natural world. It was measured by a scale used by **Roy (2015).** The scale contained statements regarding relation between stubble burning, healthy environment and various factors like economy, conservation of natural resources, climate change, etc. The responses were recorded on a three-point continuum of Agree (3), Neutral (2) and Disagree (1). After calculating their cumulative score, the respondnets were grouped into three categories namely 'low, 'medium' and 'high' by using mean (17.02) and SD (1.87).

S. No.	Ecological Consciousness	Range (Score)
1.	Low	<15
2.	Medium	15-19
3.	High	>19

3.6.13. Economic Motivation

It referred to the degree to which an individual is oriented towards achievement of the maximum economic ends such as maximization of profits. It was studied by using a scale developed by **Supe (2007)**. The scale consisted of items which covered opinions of farmers regarding producing higher yields, maximized returns/profits, trying newer and varied methods and ways to fetch incomes from different sources, etc. The responses were recorded on five-point continuum ranging from Strongly Agree (5) to Strongly Disagree (1). After calculating their cumulative score, the respondnets were grouped into three categories namely 'low, 'medium' and 'high' by using mean (21.92) and SD (2.92).

S. No.	Economic Motivation	Range (Scores)
1.	Low	<19
2.	Medium	19-25
3.	High	>25

3.6.14. Awareness about stubble management measures

It referred to the state of the respondents whether or not they have the knowledge about alternative ways to utilize the paddy stubble instead of burning. Various stubble management measures were listed after concerning literature and past studies done in Punjab and in consultation with experts. The list of alternate measures included 13 options such as charcoal production, methane production, in-situ incorporation, zero tillage/Happy Seeder, thatching, compost making, biochar production, mushroom cultivation, paper mill industry, etc. It was measured by the help of a schedule in which the respondents were awarded a score of (1) if they were aware of the measure and (0) if not.

S. No.	Awareness about stubble management measures	No. of measures respondent was aware of
1.	Low	<5
2.	Medium	5-9
3.	High	>9

The respondents were classified into three levels of awareness, viz. high, medium and low based on the number of alternative measures they were aware of. These categories were made on the basis of mean (7.02) and SD (1.54) of the total scores of the respondents. Also, the overall frequency and percentage of the respective measures was calculated to identify which measure was most known to the farmer respondents.

3.6.15. Cropping Intensity

It is the ratio of net sown area to the total cropped area cultivated by the respondent. It was measured by the following formula:

$$\text{Cropping intensity} = \frac{Grossed\ Crop\ Area}{Net\ Sown\ Area} \times 100$$

The respondents were further categorized on the basis of mean (185.81) and SD (2.86), into three categories of high, medium and low.

S. No.	Cropping Intensity	Range (%)
1.	Low	< 182.95
2.	Medium	182.95- 188.67
3.	High	>188.67

3.6.16. Attitude towards Stubble Burning

It referred to the overall evaluation of positive or negative aspects of stubble burning by the farmer. An instrument developed by Wang *et al.* (2018) was modified and used for the study. The scale consisted of six items and the responses were recorded on a five-point continuum of scores ranging from Strongly Disagree (1) to Strongly Agree (5). The scale items covered aspects like whether or not the stubble burning is the most economical stubble management measure; helpful in timely sowing of succeeding wheat crop; causes pollution; bad for human health; controls weeds and pests; and the last, whether or not it should be banned. The complete scale used is given in Annexures. A higher score meant favourable attitude towards stubble burning. Based on the responses, the respondents were categorized as following using mean (24.17) and SD (2.06).

S. No.	Attitude towards Stubble Burning	Range (Scores)
1.	Negative	<22
2.	Neutral	22-26
3.	Positive	>26

3.6.17. Subjective norms associated with stubble burning

It referred to the overall compliance with the influences of the opinions of significant others to the respondent for or against burning the paddy stubble. It indicates the motivation to comply with important people (significant others) in the respondent's life. These included fellow farmers, the farmer groups who influence the decisions of farmers related to agriculture. Scale by Wang *et al.* (2018) was modified for the study. The scale had four items and the responses were recorded on a five-point continuum of scores ranging from Strongly Disagree (1) to Strongly Agree (5). The higher score meant strong influence of the social norms associated with stubble burning. The complete scale used is given in Annexures. Based on the responses, the respondents were categorized as following using mean (12.81) and SD (2.81).

S. No.	Subjective norms associated with stubble burning	Range (Scores)
1.	Weak	<10
2.	Moderate	10-14
3.	Strong	>14

3.6.18. Perceived Behavioural Control related to stubble burning

It referred to the perceived ease or difficulty to burn the paddy stubble or perception about factors that might facilitate or inhibit the stubble burning. The scale by Wang *et al.* (2018) was used in the study. The scale had five items and the responses were recorded on a five-point continuumof scores ranging from Strongly Disagree (1) to Strongly Agree (5). The items covered aspects like the ease related to burn the stubble, monitoring, penalties, etc. The complete scale used is given in Annexures. A higher score of the respondent meant more perceived capability. i.e. more perceived ease in burning the stubble. Based on the responses, the respondents were arranged in following categories as following using mean (18.31) and SD (2.33).

S. No.	Perceived Behavioural Control related to stubble burning	Range (Scores)
1.	Low	<16
2.	Medium	16-20
3.	High	>20

3.6.19. Behavioural Intention regarding Stubble Burning

It referred to the degree of readiness of the farmers to burn or not to burn the paddy stubble after harvesting. The scale by Wang *et al.* (2018) was used in the study. The scale had three items and the responses were recorded on a five-point continuumof scores ranging from Strongly Disagree (1) to Strongly Agree (5). A higher score indicated a higher level of intention to burn the stubble. Based on the responses, the respondents were arranged in following categories as following using mean (10.22) and SD (1.85). The complete scale used is given in Annexures.

S. No.	Behavioural Intention regarding Stubble Burning	Range (Scores)
1.	Low	<8
2.	Medium	8-12
3.	High	>12

3.6.20. Stubble Burning Behaviour

The stubble burning behaviour was defined as "the action of the farmers to burn the paddy stubble on the field after harvesting and before the sowing of subsequent wheat crop." The instrument by Wang *et al.* (2018) was modified for the study. It was studied using two components or items: (i) frequency of engagement in stubble burning behaviour and (ii) the proportion of the total produced paddy stubble managed through burning. The 'frequency' referred to the number of occasions on which the farmer burnt the stubble on field after harvesting in past five years (considering one occasion per year as paddy is grown once a year in the study area). On the other hand, the 'proportion' referred to the part of the total produced stubble which was managed through burning, in case other stubble management methods were also used by the farmer.

For analysing the frequency of engagement in stubble burning behaviour by the farmer respondents, following score scheme was used:

S. No.	Frequency of engagement in Stubble Burning Behaviour since past 5 years	Score
1.	Very low (Once)	1
2.	Low (Twice)	2
3.	Medium (Thrice)	3
4.	High (Four times)	4
5.	Very High (Five times)	5

For finding out the proportion of the total stubble managed by burning, following score scheme was used:

S. No.	Proportion of the total produced stubble managed through burning	Score
1.	Very little (Less than 20%)	1
2.	Little (20% to 40%)	2
3.	Medium (40% to 60%)	3
4.	Large (60% to 80%)	4
5.	Very large (More than 80%)	5

Finally, the scores from both the components were added and a cumulative score was calculated for each respondent. A higher score indicated a higher level of performance of the respondent in terms of stubble burning. Based on the responses, the respondents were arranged in following categories using mean (7.51) and SD (1.85).

S. No.	Stubble Burning Behaviour	Range (Scores)
1.	Low	<6
2.	Medium	6-8
3.	High	≥ 8

3.6.21. Social Marketing Plan

The purpose of the present study was to analyse the stubble burning behaviour of the farmers using the Theory of Planned Behaviour (TPB). The constructs of the TPB, namely attitude, subjective norms and perceived behavioural control are basically the determinants of the behavioural intention which, in turn determines the behaviour. After estimating the relationship between the TPB constructs and finding their extent of influence on the behaviour, it became imperative to chalk out a plan that could be helpful in preventing the stubble burning behaviour of the farmers.

Social marketing is a process that applies marketing principles and techniques to create, communicate, and deliver value in order to influence target audience behaviours that benefit society (public health, safety, the environment, and communities) as well as the target audience" (Kotler, Lee and Rothschild, 2006). Social marketing has proven to be quite fruitful in inducing sustainable behavioural changes among the societies concerning different behaviours across the globe. The stubble burning relates to environment and health, to say the least, it was thus chosen to prepare a social marketing plan to tackle this behaviour. The social marketing plan for the study was formulated using BEHAVE framework given by AED (Academy for Education Development), one of the world's foremost social change organizations, working globally to improve building the foundation of thriving societies. The BEHAVE framework has been proved significantly in inducing behaviour change across varied social issues like oral rehydration therapy (ORT) to reduce infant mortality from diarrhoea, immunizations, promoting the use of condoms, TB testing services, anti-malaria bed nets, and beta-blockers. It has also helped protect watersheds, saved reefs from over-fishing, reduced small engine pollution, and helped teens protect themselves from HIV/AIDS. The list goes on- from public health to environmental protection, to traffic safety to education reforms (AED, 2008). The BEHAVE framework is based on the presumption that before any intervention, i.e. a message, a system change, an outreach effort, etc. is planned, following three questions are answered:

1. Who is the target audience, and what is important to that audience?

2. What is expected of the audience to do?

3. What are the factors or determinants that influence or could influence the behaviour, and are they determinants that a program can act upon?

   Once these three questions are answered, then the final question is to be answered,

4. What interventions could be implemented that would influence those determinants so that the determinants, in turn, could influence the behaviour?

The answers to these questions are supposed to be the steps of the BEHAVE framework (Figure 6).

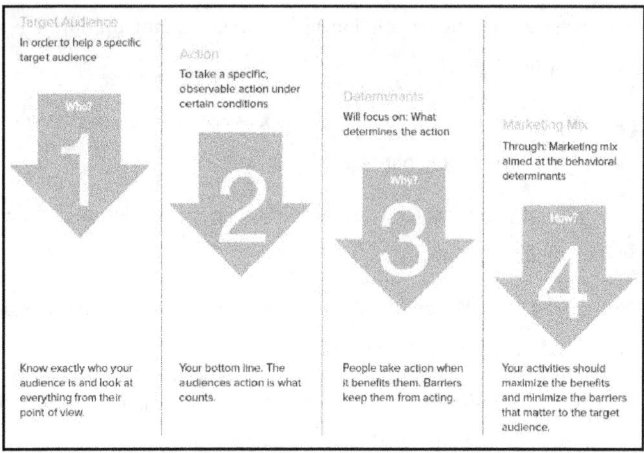

Figure 6: The BEHAVE framework used for preparing social marketing plan

The enquiries related to the feasible social marketing plan were catered by a series of open-ended questions and informal discussions considering the above BEHAVE framework. The aspects covered in these open-ended questions were framed after concerning available literature related to the study and thorough discussion with the experts.

3.7. HYPOTHESES

According to the set objectives of the study, following hypotheses were postulated:

H_{01} : There was no significant relationship between attitude of the farmers towards stubble burning and their behavioural intention regarding stubble burning.

H_{02} : There was no significant relationship between subjective norms of the farmers associated with stubble burning and their behavioural intention regarding stubble burning.

H_{03} : There was no significant relationship between farmers' perceived behavioural control related to stubble burning and their behavioural intention regarding stubble burning.

H_{04} : There was no significant relationship between behavioural intention regarding stubble burning and stubble burning behaviour of the farmers.

H_{05} : There was no significant relationship between age of the farmers and their stubble burning behaviour.

H_{06} : There was no significant relationship between size of land holding of the farmers and their stubble burning behaviour.

H_{07} : There was no significant relationship between annual income of the farmers and their stubble burning behaviour.

H_{08} : There was no significant relationship between livestock possession of the farmers and their stubble burning behaviour.

H_{09} : There was no significant relationship between information seeking behaviour of the farmers and their stubble burning behaviour.

H_{010} : There was no significant relationship between innovativeness of the farmers and their stubble burning behaviour.

H_{011} : There was no significant relationship between risk orientation of the farmers and their stubble burning behaviour.

H_{012} : There was no significant relationship between scientific orientation of the farmers and their stubble burning behaviour.

H_{013} : There was no significant relationship between ecological consciousness of the farmers and their stubble burning behaviour.

H_{014} : There was no significant relationship between economic motivation of the farmers and their stubble burning behaviour.

H_{015} : There was no significant relationship between awareness of the farmers regarding alternative stubble management measures and their stubble burning behaviour.

H_{016} : There was no significant relationship between cropping intensity and stubble burning behaviour of the farmers.

H_{017} : Gender of the farmers and their stubble burning behaviour are independent of each other.

H_{018} : Education of the farmers and their stubble burning behaviour are independent of each other.

H_{019} : Type of farming and stubble burning behaviour of the farmers are independent of each other.

3.8. DATA COLLECTION TOOLS AND TECHNIQUES

The process and procedure of data collection is an essential and significant part of any research work. The techniques for data collection comprised of specific procedure by which the researchers gather and arrange the data prior to their logical or statistical manipulation (Goode and Hatt, 1983). The study was based on information collected by primary field data as well as secondary source of data. The primary data for the current research study was collected with the help of field survey techniques. Field surveys capture snapshots of practices, beliefs, or situations from a random sample of subjects in field settings through a survey questionnaire or through a structured interview. Since the present study was of quantitative nature, most of the data was collected through a semi-structured interview schedule. The schedule represents the formal method of securing facts that are in objective form and easily described. On the basis of a thorough review of available literature, in the light of objectives set forth for the study and keeping in view, the scope and purpose of the study, a detailed and comprehensive interview schedule consisting of structured and unstructured questions was developed for collecting relevant information from the respondents. The prepared interview schedule was then consulted with experts in the field of Extension Education, Environmental Sciences and Agronomy. The suggestions made by the experts were considered in finalizing the semi-structured interview schedule. Thus, the final semi-structured interview schedule was divided into three major sections. The first section consisted of socio-personal, economic, communication, psychological and situational characteristics of the respondents. The second section constituted the constructs of Theory of Planned Behaviour namely Attitude towards stubble burning, Social norms associated with stubble burning, Perceived Behavioural Control related to stubble burning, Behavioural intention regarding stubble burning and Stubble burning behaviour. The third and final section consisted of the open-ended questions regarding the planning of social marketing plan.

3.8.1. Pilot testing of interview schedule as a data collection tool

Prior to actual collection of data and final implementation, it was necessary to test the designed interview schedule to ensure its relevance, clarity, feasibility and appropriateness. Once the tools were developed, their accuracy and appropriateness were tested through pilot testing in villages similar to the study villages. Pre-testing of interview schedule helped in determining whether the respondents had any difficulty in responding to the statements and questions included in the schedule. Another reason for carrying out pilot testing was to check the feasibility of the selection of research problem for the present investigation. The pre-testing of interview schedule was carried out by interviewing 30 non-sampled respondents of two villages namely Bibipur and Jatan of Khanna block in Ludhiana district. At the time of pre-testing, the purpose of interview and study was explained to the farmers. Pilot testing was carried out for the duration of about fifteen days from 6th September, 2019 to 20th September, 2019. During the period of pilot testing, the researcher had a stay in house of a known relative working as a Territory Manager in the company named Mahindra Tractors in Ludhiana. Another purpose of the pilot-testing was to build rapport among the villagers and to get acquainted with the regular village lifestyle of people in these villages.

3.8.2. Collection of data

On the basis of pre-testing of the appropriateness of the developed tool, necessary modifications, wherever needed were incorporated in the selection and sequencing of statements and questions. The finalized schedule was then administered to the sampled respondents and data was collected through personal interview method. The final data for the study were collected by interviewing total 246 farmers of the selected villages just before the starting of the paddy harvesting season in the first week of October 2019 to mid-December 2019. The respondents were contacted personally in an informal way at their fields, offices, shops or at their residence. Before conducting the interview, the purpose and objectives of the investigation were explained to the respondents by the researcher to obtain correct and wholehearted answers from them. Each question was explained to the farmer respondents with equal emphasis. The informal discussions and observations were also utilized as a part of investigation in order to understand the respondents and their situation.

Statistical facts and information relevant to the present investigation in the form of statistical abstracts, annual reports, progress reports, village directories, census abstracts, etc. were collected from the official websites of government and non- governmental organizations. Besides, secondary data from various published and unpublished literatures were also used to acquire necessary information.

3.9. TYPE OF DOCUMENTATION

In order to record data at the time of observation, interview and discussion without loosing its essence, different types of records were maintained as below. These were used to document daily account of activities, experiences, people, places and events simultaneously.

3.9.1. Log Book: A small note book was used to immediately jot down the emotional account of the observation on the spot. It was written at the end of the day about his experience by critically observing the village life. It was done to avoid any possible error caused by over dependence on memory.

3.9.2. Field Dairy: At the end of the day, the researcher used to write a complete description of her activities for that day. This helped in tracing out the occurrence of any particular event by critically observing the whole activities and events.

3.9.3. Photographs: Photographs were used to document people, places, events and setting over time. This helped in describing the real-life situation of the villagers, creating visual impact through photographic record of special behaviour. Appropriate photographs in various places were helpful in the documentation process as a matter of strong evidence.

3.9.4. Voice Recorder: Voice recorder was used to record relevant conversation among the researcher and the respondent. This was done with due permission from the informants. It was also used to capture the evidences of the actual words spoken at the time of focus group discussions. It served as the authentic matter for analysis and documentation.

3.10.   STATISTICAL ANALYSIS AND INTERPRETATION OF DATA

Statistical analysis of quantitative data is an important aspect of research work, as it facilitates the condensation and interpretation of collected data in simple form and helps in predicting trends. Statistical Package for the Social Sciences (SPSS, ver. 21) software was used for analysing, tabulating and interpreting data in the light of the objectives set forth for the study. The statistical techniques used for data analysis were as follows:

3.10.1. Frequency

It was calculated to find out the number of respondents in a particular cell and was used to categorize the respondents into various categories on the basis of selected variables.

Freshly harvested paddy field with left-over stubble

Sowing of wheat crop on stubble-burnt field

Burning of the left-over paddy stubble

Square-shaped bales of paddy stubble being transported

Researcher interacting with farmer respondents during data-collection process

Plate 3: Excerpts from field visits in the study area

3.10.2. Percentage

Percentage values were calculated to make simple comparisons. These were calculated by dividing the frequency of a particular cell by total number of respondents and multiplying by 100.

P = (n/N) ×100

Where,

P = Percentage

n = Frequency of a particular cell

N = Total number of sample respondents

3.10.3. Arithmetic mean

Mean score for each category were worked out separately by the formula.

$$\bar{x} = \frac{\sum x_{ij}}{n}$$

Where,

X_{ij} = Sum of each of the individual comparisons

I = 1, 2, 3….. n

n = sample size

$\overline{x}$ = Arithmetic Mean

3.10.4. Standard Deviation

It is a measure of variability in a set of scores, found by summing up the squared differences from the mean, dividing by the number of cases and extracting the square root. It was computed for the purpose of analysis and further categorization. The formula was used for calculating standard deviation was as follows.

$$s = \sqrt{\frac{\sum (x - \bar{x})^2}{n - 1}}$$

Where,

     s   =   Standard Deviation

     x   =   Value of individual Variable

     $\bar{x}$ =   sample mean

     n   =   sample size

3.10.5. Weighted Mean Score

Weighted mean score was calculated by using the following formula:

$$a_w = \frac{\sum mW}{W}$$

Where,

a_w = Weighted mean score

mW = Product of weight and measurement

W = Total of observation

3.10.6. Co-efficient of correlation

Correlation coefficient is a number computed from a set of data which summarizes the extent to which variations in one measure go together with variation in other measures. It was used to find out the relationship between independent and dependent variables. The formula used to compute the co- efficient of correlation is as follows:

$$r = \frac{\text{Cov.}(x_1 x_2)}{\sqrt{\text{Var}(x_1)\,\text{var}(x_2)}}$$

Where,

r = Correlation coefficient

X_1 and X_2 are the two variables

Var $X_1 = \sum(x_1 - \bar{x}_1)^2/n$

Var $X_2 = \sum(x_2 - \bar{x}_2)^2/n$

Covar. $(x_1 x_2) = \sum(x_1 - \bar{x}_1)(x_2 - \bar{x}_2)/n$

3.10.7. Chi-square test

As a test of independence, chi-square (χ^2) test enables to explain whether or not two variables are associated. For the present study, it was used to know whether the profile characteristics with grouped data were dependent on the stubble burning behaviour of the farmers or not.

First of all, the null hypothesis that the two variables are independent was created. The expected frequencies were calculated and then the value of χ^2 was calculated as follows:

$$\chi^2 = \sum_{i=1}^{r} \sum_{j=1}^{c} \frac{(O_{i,j} - E_{i,j})^2}{E_{i,j}}$$

Where,

Oij = observed frequency of the cell in ith row and jth column.

Eij = expected frequency of the cell in ith row and jth column.

If the calculated value of χ^2 is less than the table value at 5% level of significance for given degrees of freedom, it is concluded that null hypothesis stands which means that the two attributes are independent of each other. But if the calculated value of χ^2 is greater than its table value, the inference then would be that null hypothesis does not hold good which means the two variables are dependent on each other and the relation is not because of some chance factor but it exists in reality.

3.10.8. Structural Equation Modeling

Structural Equation Modeling (SEM) is a multivariate technique combining aspects of path analysis and multiple regression that enables the researcher to simultaneously examine a series of interrelated dependence relationships among the measured variables and latent constructs as well as between several latent constructs (Hair Jr. *et al.*, 2006). Relationships as hypothesised by the research model, also known as Measurement Model were tested through Structural Equation Modelling using IBM SPSS AMOS software. Structural equation modeling was used because farmers' intention and its psychological determinants are latent variables, i.e., the variables which exist but cannot be directly observed. To conduct quantitative research on practical problems like the present study, the SEM evaluates the theoretical model according to the extent of consistency between the theoretical model and the actual data. A model is representation of a theory. Theory can be thought of as a systematic set of relationships providing a consistent and comprehensive explanation of phenomena. SEM model consists of the measurement model and the structural model.

Measurement Model: It is the theoretical model. It depicts how the observed (measured) variables represent the constructs. The hypothesized relationships between latent constructs which are taken for explaining the behaviour in context are derived from a pre-existing theory. Then it is checked whether those latent constructs can explain the behaviour in question or not. In the present study, the latent constructs to explain the Stubble Burning Behaviour were derived from the Theory of Planned Behaviour and the model fit was tested using Confirmatory Factor Analysis.

Structural Model: Once the measurement model is found to have overall good fit, then the Structcural Model is specified. This model explains the observed covariance among a set of measured variables by estimating the observed covariance matrix with an estimated covariance matrix constructed based on the estimated relationships among variables. The closer the estimated and observed covariances are, the better is the model fitness. In other words, a structural model shows how the constructs are interrelated to each other, often with multidependence relationships. It specifies whether a relationship between the variables exists or not.

According to Nayak (2017), the overall procedure of Structural Equation Modelling consists of six stages. The first four stages correspond to assessing the measurement model using Confirmatory Factor Analysis while the last two stages correspond to assessing the structural model using path analysis.

Stage 1: Defining the individual constructs- The process begins by listing and defning the constructs that comprise the measurement model. This process was completed in the previous sections where the constructs related to Theory of Planned Behaviour were defined.

Stage 2: Developing the overall measurement model- In this stage, all the individual constructs were brought together to form an overall measurement model.

This stage and the stages following used AMOS software for carrying out the analysis.

Stage 3: Testing the measurement model- In this stage, the measurement model was tested. All the standard rules and procedures were followed.

Stage 4: Assessing the measurement model fit- The measurement model fit was assessed at this stage. The fit compares the observed and expected covariance matrices. Different standard measures of overall model fit include Absolute, Incremental, and Parsimonious indices of fit. Seven conventional model-fit statistics were chosen for the present study. The best-known index of Absolute fit is the Chi-square which is denoted by χ^2/degrees-of-freedom (χ^2/df). The Goodness of Fit Index (GFI) and the Comparative Fit Index (CFI) were the two indices used to evaluate the overall absolute fit of the proposed model. To evaluate the incremental fit measures, the Adjusted Goodness of Fit Index (AGFI), the Incremental Fit Index (IFI), and the Normed Fit Index (NFI) were used. Finally, the Root Mean Square Error of Approximation (RMSEA) was used to evaluate the parsimonious fitness of the proposed model.

Stage 5: Specifying the structural model- this stage involves specifying the structural model by assigning relationships from one construct to another based on the proposed theoritica model. Structural model specification focuses on adding single-headed, directional arrows to represent structural hypothesis in the resarcher's model. In other words, the dependence relationships that are hypothesized to exist among the constructs are identified.

Stage 6: Assessing the structural model fit- The final stage involves testing the fitness of the structural model and its corresponding hypothesized theoretical relationships. When the measurement model is validated and achieves acceptable model fit then only the structural model can be tested. Similar fit indices were used for the structural model as well. After that the path estimates were gathered and analysed for checking the existence of relationships between TPB constructs.

Results
and
Discussion

This chapter deals with the research findings of the present investigation along with their relevant discussion. The data obtained from respondents were measured, analysed, tabulated and statistically treated according to the objectives of the study set forth. The facts emerged from statistical treatments of the data have been presented and discussed under following subheads:

4.1   Profile characteristics of the farmers

4.2   Stubble burning behaviour of the farmers

4.3   Relationship between Theory of Planned Behaviour Constructs

     4.3.1.   The Measurement Model

     4.3.2.   The Structural Model

     4.3.3.   Extent of influence of attitude, subjective norms, perceived behavioural control on behavioural intention regarding stubble burning

     4.3.4.   Extent of influence of behavioural intention on stubble burning behaviour of farmers

4.4   Relationship between profile characteristics and stubble burning behaviour of the farmers

4.5   Social marketing plan

4.1. PROFILE CHARACTERISTICS OF THE FARMERS

The first and foremost thing to be addressed by the researcher during a survey research study is the profile of its respondents. This paves the way for further scientific treatment of data according to the set objectives. During the course of the present study, an attempt has been made to understand various socio-personal, economic, communication, psychological and situational characteristics of the farmers in the study area. These characteristics included age, gender, education, size of land holding, annual income, livestock possession, type of farming, information seeking behaviour, innovativeness, risk preference, scientific orientation, ecological consciousness, economic motivation, awareness about stubble burning measures and cropping intensity. These characteristics were investigated using appropriate measurement procedures and the results of the findings have been presented under the following sub heads:

4.1.1. Age

Age is given due consideration as a significant variable with respect to experience and responsibility. It can play a prominent role in determining the level of social and economic participation of an individual in a social system. Data regarding age composition of respondents has been presented in Table 8. It reveals that majority of the respondents (65.46%) residing in the sampled villages belonged to the middle age category (26 to 50 years). There were only 19.10 per cent farmer respondents who belonged to the old age category (>50 years) followed by 15.44 per cent of the farmers who were in the young age category (<26 years) respectively.

Table 8: Distribution of respondents on the basis of age (n=246)

S. No.	Age	Frequency	Percentage
1	Young (less than 26 years)	38	15.44
2	Middle-aged (In between 26 and 50 years)	161	65.46
3	Old (more than 50 years)	47	19.10

The findings of the current investigation indicate that most of the farmer respondents belonged to the middle age category, i.e. within the age-group of 26-50 years. While the least number of farmers during the study were found to be in the young age category, i.e. less than 26 years of age. The reason behind this might be the growing reluctance of farm families to indulge the upcoming generations in agriculture as a traditional or family profession due to decreasing profits in the venture. On the other hand, the majority of people from young age-group were also not found to be much interested in continuing agriculture or taking up other allied venture as a form of business. Instead, they were keenly interested in going abroad particularly Canada, Australia, New Zealand and United Kingdom and settling down there. The parents are willing to send their children abroad even if they have to take up some kind of loan. As a consequence, most of the respondents belonging to middle and old age groups happen to be the farmers who are left behind. So, the lack of interest among majority of the younger generations in taking up agriculture as an occupation, might have contributed to the higher percentage of farmers in middle aged category in the sampled villages.

The results of the present study are in line with the studies conducted by **Prasad (2002), Tamizhkumaran and Rao (2012), Verma (2015), Parganiha and Sharma (2017), Mandal (2018), Saklani (2018), Palvi *et al.* (2018)** who reported that most of the farmers belonged to the category of middle age group. The results of the studies conducted by **Anand (2016), Singh *et al.* (2016), Sharma *et al.* (2017) and Lyngdoh (2018)** also revealed that most of the farmers in Punjab were found in the middle-aged category.

4.1.2. Gender

The gender distribution of the respondents is presented in Table 9. It is evident from the data that majority of the respondents were men (91.06%) and only 8.94 per cent of the total farmer respondents were women.

Table 9: Distribution of respondents on the basis of gender (n=246)

S. No.	Gender	Frequency	Percentage
1.	Men	224	91.06
2.	Women	22	8.94

This situation is due to the fact that in the study area, most of the agricultural activities are carried out by men and women are primarily engaged in household and other activities. Most of the women respondents took up farming as their primary occupation under unfortunate family situations such as passing away of the male head of the family. Although, there were a few women farmer respondents who took up farming by their own choice after pursuing higher studies in agriculture. They were willing to apply the learning of their academics to practical fields. This also shows a changing progressive trend in which women are also coming up with to take up agriculture and allied activities as their professional ventures on their own. Although, the numbers still did not reflect the "trend" per se, as for now, but a slight deviation from the existing patterns has certainly started when it comes to bearing farming responsibilities. Studies including **Indimuli (2013), Imarhiagbe (2015)** and **Joshi (2016)** also found that majority of farmers were men in their respective researches.

4.1.3. Education

The educational status of respondents has been presented in Table 10. It shows that maximum percentages of the respondents (21.95%) had education up to Intermediate level followed by 21.54 per cent respondents belonged to graduate level of education. Among all the respondents, 19.10 per cent had formal education up to high school level. Around 13 per cent of the farmer respondents had received formal primary education while 12.20 per cent farmers were found to be functionally literate. Only 5.29 per cent of the farmer respondents were Post Graduates while 6.91 per cent farmers were illiterate.

Table 10: Distribution of respondents on the basis of education (n=246)

S. No.	Education	Frequency	Percentage
1	Illiterate	17	6.91
2	Functionally literate	30	12.20
3	Primary education	32	13.01
4	High school	47	19.10
5	Intermediate	54	21.95
6	Graduate	53	21.54
7	Post graduate	13	5.29

The data in Table 10 further indicates that a considerable proportion of farmer respondents (19.11%) were found to be either illiterate or functionally literate. Most of these respondents belonged to the old-age group who basically lacked enthusiasm, need and resources to access formal education during their times. On the other hand, more than half (54.06%) of the total farmer respondents couldn't or didn't continue their formal education beyond school level. It included respondents who formally studied up to primary level, high school level and intermediate level. It was also noted that majority of these respondents belonged to middle-age group who had a limited access to resources and had identified education as a basic need to some extent. The results further revealed that more than a quarter (26.83%) of the farmer respondents had pursued higher education, i.e. Graduation and Post-Graduation. Majority of this chunk of respondents belonged to young-age group who had access to all kinds of resources in their close proximity to pursue their educational goals. So, it was noted that in the selected villages, the need of having a formal education has emerged over a period of time, and with time, have emerged the opportunities, resources and facilities. Currently, all the sampled villages had a primary school and were also well linked with the nearby towns which had private and government schools and colleges. The current generation parents were found to be serious regarding the studies of their children and were availing the opportunities and resources to educate their children as per their best convenience. Thus, it can be said that although literacy rate varies from place to place but people have become serious about education. As of today's generation, although not many people are opting for higher education but still basic school education was found to be common to all. Thus, it can be concluded from the above research findings that farmers have recognized education as a basic need over a period of time and are trying their best to provide the same to budding generations in the best way possible. This also marks the beginning of the base formation a strong educated society.

Similar results were noted in the studies of **Verma (2015), Parganiha and Sharma (2017), Palvi *et al.* (2018)** and **Saklani (2018)** in which the maximum number of farmer respondents had intermediate or higher/senior secondary level of formal education. Studies conducted by **Anand (2016), Singh *et al.* (2016)** and **Sharma *et al.* (2017)** in the state of Punjab also found that the majority of farmers had completed Intermediate and medium level of formal education in their respective studies.

4.1.4. Size of land holding

The distribution of farmer respondents on the basis of their size of land holding is depicted through Table 11. It is evident from data that the maximum number of respondents (27.24%) possessed semi-medium size, i.e. 5-10 acres of landholding. Farmers with small-sized landholding (2.5-5 acres) accounted for 21.95 per cent, which were followed by the farmers possessing medium sized land (19.51%), i.e. 10-25 acres. The marginal size of land holding (less than 2.5 acres) was held by 18.70 per cent of the farmer respondents, while 12.60 per cent of the farmers owned large size landholdings, i.e. more than 25 acres.

Table 11: Distribution of respondents on the basis of size of land holding (n=246)

S. No.	Size of Land holding	Frequency	Percentage
1	Marginal (less than 2.5 acres)	46	18.70
2	Small (In between 2.5 and 5 acres)	54	21.95
3	Semi-medium (In between 5 and 10 acres)	67	27.24
4	Medium (In between 10 and 25 acres)	48	19.51
5	Large (more than 25 acres)	31	12.60

According to **Punjab Economic Survey, 2019-20,** the maximum number of land holdings in Punjab is categorized under semi-medium category (34%) followed by medium (28%), small (19%), marginal (14%) and only five per cent under 'large' categories respectively. So, the findings of the present investigation are almost parallel to the state figures in terms of proportion.

The above findings are found to be in line with those of **Mohapatra (2013), Anand (2016), Lyngdoh (2018)** and **Singh** *et al.* **(2019)** which reported that the highest number of farmer respondents had semi-medium or medium sized land holdings. On the other hand, the current findings stand contradictory to those of **Joshi (2016), Mandal (2018)** and **Saklani (2018)** which reported that maximum farmer respondents had marginal-sized land holdings.

4.1.5. Annual Income

The data regarding annual income is presented in Table 12 which clearly indicates that the majority of the farmers (51.63%) were categorized under 'low' annual income category. The respondents under this category earned in between Rs. 50,000 and Rs. 4,50,000 annually. The farmers who were classified under 'medium'

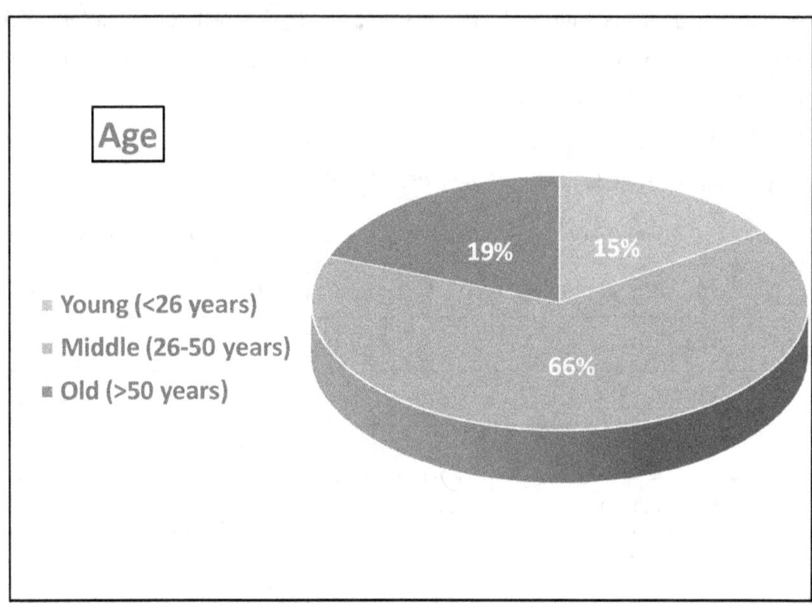

Figure 7: Distribution of respondents according to age

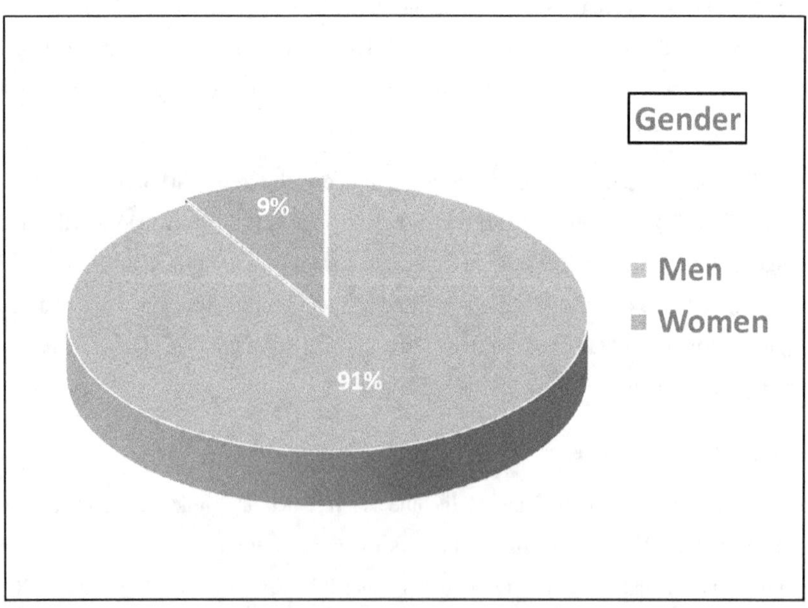

Figure 8: Distribution of respondents according to gender

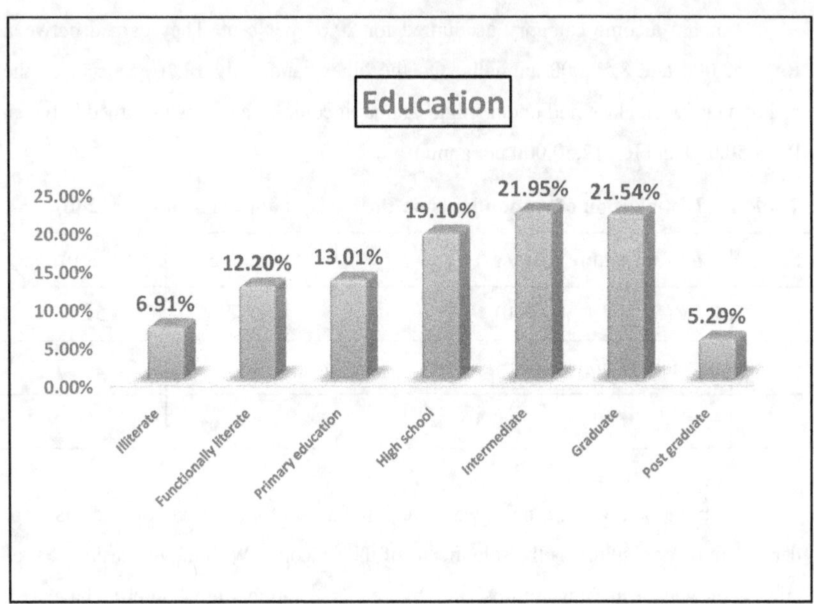

Figure 9: Distribution of respondents according to education

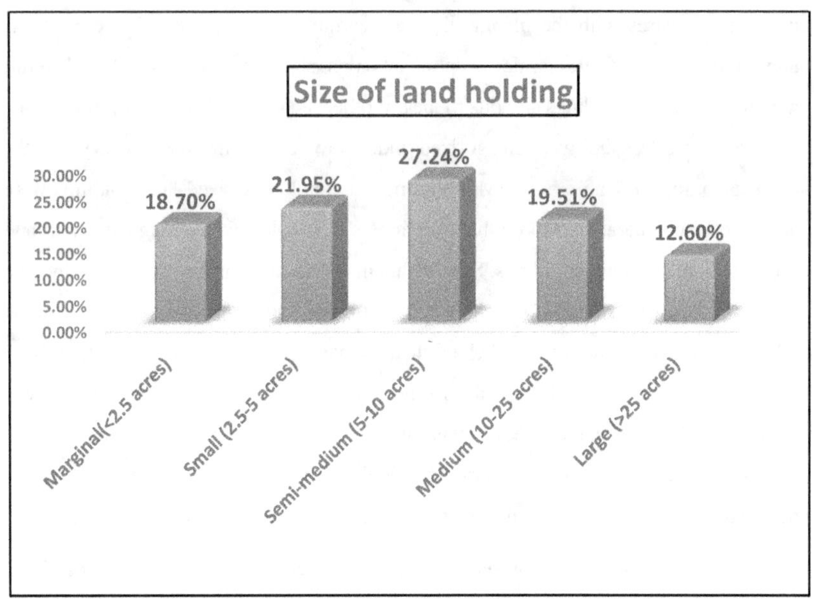

Figure 10: Distribution of respondents according to size of land holding

annual income category accounted for 29.67 per cent. They earned between Rs. 4,50,000 and 8,50,000 annually. On the other hand, only 18.70 per cent of the respondents were classified under 'high' annual income category who earned between Rs. 8,50,000 and Rs. 12,50,000 per annum.

Table 12: Distribution of respondents on the basis of annual income (n=246)

S. No.	Annual Income (Rupees)	Frequency	Percentage
1	Low (50,000 - 4,50,000)	127	51.63
2	Medium (4,50,000 - 8,50,000)	73	29.67
3	High (8,50,000 - 12,50,000)	46	18.70

During the investigation, it was found that most of the farmer respondents were dependent on agriculture as the sole means of their income. With the increasing cost of cultivation and decreasing returns year by year, the farmers are bound to have less income. Some of the small and marginal farmers even expressed their concern for agriculture in the upcoming future because it is becoming hard for them to sustain and meet the expenses with the income they are earning from the limited resources. As around two-third of the farmer respondents possessed marginal, small and semi-medium sized land holdings and due to lack of other secondary occupations, most of the farmers earned less and were categorized under 'low' annual income category. On the other hand, most of the farmers who hold medium and large sized land holdings were able to earn comparatively more than the famers who belonged to marginal, small and semi-medium sized land holdings. Some of them even had secondary sources of income like small businesses like *kirana* shops, confectionary shops, general stores, etc. in the villages and nearby towns which helped them manage their expenses. Hence, they were classified under 'medium' annual income category. The farmer respondents who belonged to 'high' income category were mostly large-sized land holders having other side businesses as well like dairy and poultry farms, *aadats* in *mandis*, tractor dealerships, cloth and car showrooms, seed plants and warehouses, etc.

The findings of the present investigation were in line with those of **Joshi (2016)** who also reported that the majority of the farmers in Uttar Pradesh belonged to 'low' income category. On the other hand, the findings of **Verma (2015), Palvi *et al.* (2018)** and **Saklani (2018)** reported that maximum number of farmers belonged to 'medium' income category, and thus, stand contradictory to the findings of present investigation. The findings of the studies done in Punjab including **Roy (2015)** and **Deepika (2017)** were found to be in line with the present study as they also reported that the majority of farmers belonged to 'low' income categories. Contradictorily, other studies done in Punjab like **Sharma *et al.* (2017)** and **Lyngdoh (2018)** reported that majority of farmers belonged to 'medium' income group.

4.1.6. Livestock Possession

Livestock forms an integral component of rural households in India. It is widely considered as an indicator of economic status of the family in a rural setting. The findings regarding livestock possession of the farmers have been presented in Table 13.

Table 13: Distribution of respondents on the basis of livestock possession (n=246)

S. No.	Livestock Possession	Frequency	Percentage
1	Poultry birds	20	8.13
2	Smaller milch animals (Sheep or Goat)	9	3.66
3	Larger milch animals (Cow or buffalo)	243	98.78
4	Draught animals (Oxen)	11	4.47

***Multiple responses were allowed**

The data in Table 13 reveals that almost all the respondents (98.78%) possessed larger milch animals including buffaloes and cows, especially buffaloes. On the other hand, poultry animals were kept by 8.1 per cent of the respondents while 4.47 per cent of the total respondents possessed draught animals like oxen. The findings also show that smaller milch animals like sheep or goat were reared by just 3.66 percent of the farm respondents.

During the investigation, it was noted that almost all the farm families had at least one milch animal especially buffalo because the people in villages preferred consuming buffalo milk over any other type of milk. Several households possessed

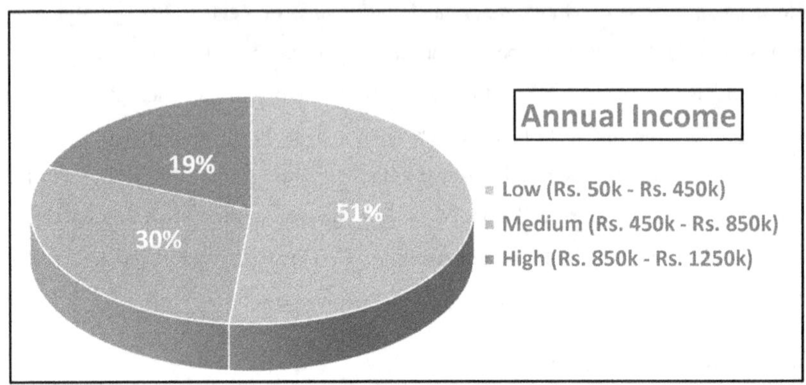

Figure 11: Distribution of respondents according to annual income

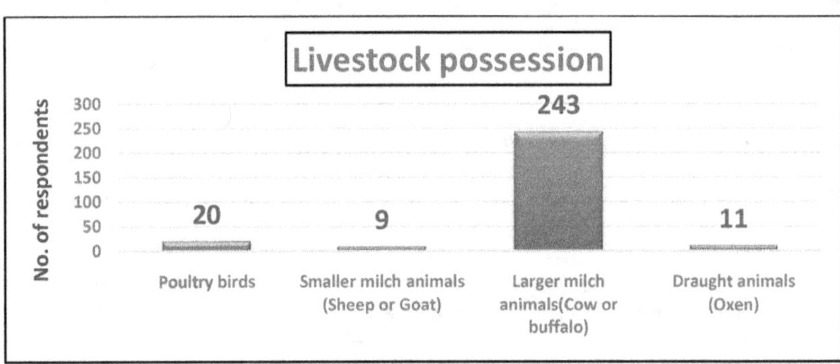

Figure 12: Distribution of respondents according to livestock possession

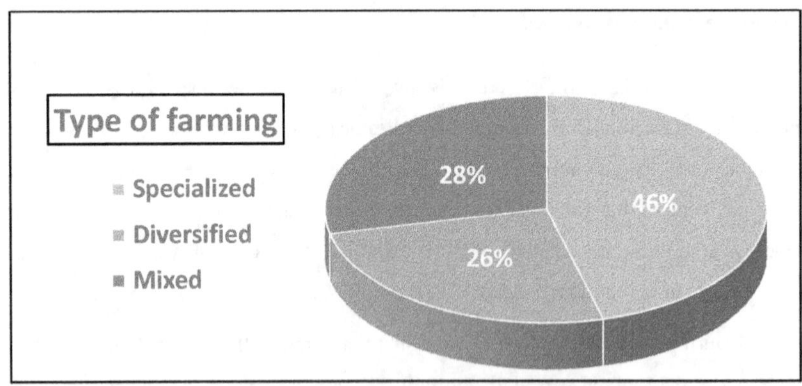

Figure 13: Distribution of respondents according to type of farming

cows as well. Some of the farmers, who had excess production of milk, they sold it in nearby towns also. Some of the farmer respondents had poultry farms as a subsidiary venture and thus, kept poultry birds. Only a few households had kept goat because the majority of the people had larger milch animals for milk. The goat milk was used mostly for medical purposes. It also fetched good price to the farmers. People from nearby towns also used to come and get it in case of emergency. Although, mechanization has taken over the manual or animal drawn operations in agriculture in most of the Punjab, some farmers had still kept oxen. Occasionally, farmers earned well from the semen of oxen which was used in artificial insemination of cattle.

Similar findings were reported in the studies of **Karuna (2013), Basera (2019)** in Uttarakhand and **Dash (2019)** in Odisha, who also found that majority of the respondents possessed larger milch animals.

4.1.7. Type of Farming

The farmers were categorized on the basis of three types of farming, namely, specialized, diversified and mixed. The Data regarding type of farming practiced by the respondents has been presented in the Table 14.

Table 14: Distribution of respondents on the basis of type of farming (n=246)

S. No.	Type of Farming	Frequency	Percentage
1	Specialized	113	45.93
2	Diversified	63	25.61
3	Mixed	70	28.46

It is clear from the data that maximum number of the respondents (45.93%) practiced specialized farming. On the other hand, mixed and diversified types of farming were found to be practiced by 28.46 per cent and 25.61 per cent of the farmer respondents.

The major chunk of the farmer respondents was practicing specialized farming. It can be explained by the fact that most of the farmers were dependent solely on agriculture as a means of income; hence they grew crops on a commercial level and sold their crops in the nearby *mandis*. Such farmers had basically one motive only, i.e. to have the utmost production and yields and ultimately, which could result in maximizing profits. For that purpose, they used all kinds of inputs such as chemical fertilizers, insecticides, pesticides, growth enhancers etc. which provided better yields.

On the other hand, the farmers who were practicing mixed farming had either a dairy or a poultry farm attached to their crop fields. They had basically two sources of income, one from the field crops grown and the other from the dairy or poultry farms. Those who had cattle, sold milk and milk products like *paneer, khoya,* and sweets made by milk, etc. on their local shops. The farmers who had poultry sold birds and eggs as per their capacity. Thus, they had an additional source of income apart from the selling of the field crops in the nearby *mandis*. They were also found to be using cattle dung, or poultry litter into the fields as a source of manure.

The farmers practicing diversified farming basically had more than two alternatives on the same farm. For example, the farmers growing field crops, running a poultry or dairy farm and having a small shed for mushroom cultivation; or the farmer taking simultaneously two crops (such as intercropping) along with earning from dairy or poultry and having a small pond on the farm for fish cultivation or running both poultry and dairy farms along with field cultivation of crops. This ensured diverse alternatives of income to the farmers. Some of the respondents practicing diversified farming were cultivating crops for their own consumption rather than selling into the markets. The informal discussions with such farmers revealed that they cultivated crops with the minimal or no use of chemicals and applied biological measures for preventing disease and pest attacks. Although, not using chemical inputs results in substantially lesser yields but they were least bothered about that as the expenses were met with the other diversified sources. They believed that the amount of chemical residue in the crops is slowly but steadily affecting the health of people in an adverse manner. Hence, at least for their own home consumption, they grew crops with minimum chemical use. Apart from that, looking at the growing awareness of people and demand of chemical free or organic crop products, some of these farmers practicing diversified farming currently were trying to get into value addition and marketing of such products.

The studies including **Chauhan (2011), Meena (2012)** and **Chauhan (2018)** were found to contradict with the findings of present investigation. The above-mentioned studies reported that majority of the farmers in Almora, U.S. Nagar and Dehradun districts of Uttarakhand respectively were practicing mixed farming. Another contradictory study **Mandal (2018)** revealed that maximum farmer respondents in West Bengal were following diversified farming.

4.1.8. Information seeking behaviour

In the present study, information seeking behaviour of the farmer respondents was operationalized as the frequency of contact or exposure of the respondent to different sources viz. personal localite, personal cosmopolite, mass media and extension methods for obtaining information on agriculture and related aspects. The information seeking behaviour of the farmers was studied under four categories described below:

4.1.8.1. Information seeking behaviour from personal localite sources

The data in Table 15 represents the distribution of respondents according to their frequency of seeking information from personal localite sources. It can be inferred from the data that among the personal localite sources of information, majority of the respondents (55.70%) sought information from relatives on occasional basis with weighted mean score (WMS) of 2.06 and rank I. It was also found that 52.44 per cent of the farmers sought information from progressive farmers on the occasional basis which stood II with WMS of 2.04. The results further indicate that neighbours, with WMS 1.95 were the third most frequently utilized sources of agricultural information in the study area with 46.34 per cent of the farmers depending on them occasionally. The friends were the second least with 12.60 per cent regular usage with WMS of 1.69 and the local leaders were the least preferred information source with WMS of 1.60 and just 6.50 per cent regular information seeking by the farmers in the study area.

Table 15: Distribution of respondents according to frequency of seeking information from personal localite sources (n=246)

S. No.	Personal localite sources	Seeking behaviour						Weighted mean score (WMS)	Rank
		Regular		Occasionally		Never			
		f	%	f	%	f	%		
1	Friends	31	12.60	110	44.72	105	42.68	1.69	IV
2	Neighbours	60	24.40	114	46.34	72	29.26	1.95	III
3	Relatives	62	25.20	137	55.70	47	19.10	2.06	I
4	Progressive farmers	64	26.02	129	52.44	53	21.54	2.04	II
5	Local leaders	16	6.50	118	47.97	112	45.53	1.60	V

***Multiple responses were allowed**

It was observed during the investigation that some farmers had purchased some farm implements on shared basis with their relatives. So, the majority of information sought from relatives was based on type of farm machinery to be utilized for various operations. This included the information regarding the time, method and frequency of use of farm implements. The second most frequent information source was the progressive farmers in the study area. This shows that the farmers had more trust on the information supplied by someone of their own kind, i.e. by a farmer himself. It was noticed that the progressive farmers had higher degree of cosmopolite nature than the rest of the farmers in terms of exchange of agricultural information. The farmer respondents sought information from progressive farmers regarding new varieties, techniques of stubble management, etc. The neighbours were also rated high as the sources of agricultural information. It was noticed that neighbouring farmers took advice and discussed issues like time of sowing, irrigation and harvesting, etc. among each other. In some villages, it was noticed that the neighbouring farmers had a good understanding among themselves and rented farm machines like harvesters from the common vendor, which ultimately saved some cost.

It was further revealed that the farmers didn't consult their friends on a regular basis because most of their friends were living in distant villages. But they did share information with them on an occasional basis. The local leaders were clearly the least preferred source of agricultural information among the personal localite sources for the farmer respondents. Most of the respondents expressed their sheer displeasure towards the service and competency of the local leaders in the study area. Thus, they were reluctant to regularly consult the local leaders regarding agricultural information per se. Therefore, it might be concluded from the above findings that for seeking agricultural information, the farmers were not completely dependent on any of the single personal localite sources for agricultural information on a regular basis. Instead, they sought occasional guidance on various agricultural issues from different sources.

Data regarding information seeking behaviour from personal localite sources is presented in Table 16. The data shows that majority of the respondents (70.73%) had medium information seeking behaviour from personal localite sources followed by 25.20 per cent of respondents who had high information seeking behaviour from personal localite sources. The remaining 4.07 per cent of respondents showed low level of information seeking behaviour from personal localite sources. From the data in Table 15, it may be concluded that the respondents either showed medium or high level of information behaviour from localite sources like relatives, progressive farmers and neighbours. This might be due to the fact that these sources are in the proximity of the farmers' locality which ensured easy and timely access in the time of need.

Table 16: Distribution of respondents on the basis of information seeking behaviour from personal localite sources (n=246)

S. No.	Personal localite	Frequency	Percentage
1.	Low (Less than 7)	10	4.07
2.	Medium (In between 7 and 11)	174	70.73
3.	High (More than 11)	62	25.20

The findings reported by **Raghuvanshi (2015)** in a study expressed similar observations and revealed that about 46.38 per cent of respondents showed medium level of information seeking behaviour from personal localite sources followed by 40.90 per cent of respondents who had high information seeking behaviour from personal localite sources. **Verma and Sharma (2015)** and **Basera (2019)** also recorded similar outcomes in their research study.

4.1.8.2. Information seeking behaviour from personal cosmopolite sources:

The distribution of respondents according to the frequency of seeking information from personal cosmopolite sources is presented in Table 17. The data indicates that among personal cosmopolite sources, the Input retail shops were most frequently used by the respondents for seeking agricultural information, with weighted mean score (WMS) of 2.30 and corresponding rank I. Company Agents from various input companies like those of seeds, chemicals etc. were the second most frequently used information source with WMS of 1.63. The third most popular personal cosmopolite source of information was found to be the Field Officer from the banks with a WMS of 1.75, followed by KVK-SMS, Agricultural Officer and Block Development Officer respectively with weighted mean scores of 1.64, 1.35 and 1.32 respectively.

Table 17: Distribution of respondents according to frequency of seeking information from personal cosmopolite sources (n=246)

S. No.	Personal cosmopolite sources	Seeking behaviour						Weighted mean score (WMS)	Rank
		Regular		Occasionally		Never			
		f	%	f	%	f	%		
1	Agriculture Officer (AO)	0	0	87	35.37	159	64.63	1.35	V
2	Block Dev. Officer (BDO)	0	0	79	32.11	167	67.89	1.32	VI
3	Field Officer (Bank)	38	15.45	110	44.71	98	39.84	1.75	III
4	KVK-Subject Matter Specialists (SMS)	21	8.54	117	47.56	108	43.90	1.64	IV
5	Input Retail Shops	95	38.62	130	52.85	21	8.53	2.30	I
6	Company Agents	43	17.48	106	43.09	97	39.43	1.78	II

***Multiple responses were allowed**

The results show that the input retailers and company agents of various input companies were the most preferred sources of the information among personal cosmopolite sources. The reason behind this might be that most of the farmers in study area were using various kinds of inputs like seeds, fertilizers, chemicals for pest and insect control etc. for which, they visited the input dealers for purchasing the inputs frequently. It was revealed that during these visits to the input retail shops, farmers used to describe the condition of their standing crops to the shopkeepers and accordingly, they recommended the inputs to be used on the fields. The recommendations ranged from the type of chemical, brand of chemical and the dosage to the frequency of usage of those inputs. Thus, farmers got a wide range of information from a single point of contact. On the other hand, Punjab being a very competitive market for all kinds of farm inputs, from chemicals to farm machinery, the companies have appointed agents at various cadres to establish direct contact with the farmers for facilitating their sales.

The second most frequently used sources were these company agents. They used to contact the farmers and provided them information about latest products they had to offer and sometimes, also delivered those products at the farmers' doorsteps. Farmers also utilized that information to compare the quality and prices of the products with other agents and input retailers. The field officers from the banks had a similar approach of establishing the rapport and contact with the farmers but their information was often limited to loans and credit.

It was noted that the private-sector sources were more frequently utilized as compared to the public-sector sources for gathering farm related information by farmers in the study area. The results showed that Subject Matter Specialists from KVKs, Agricultural Officer and Block Development Officer were least consulted. The farmers revealed that these offices/KVKs were distantly located from the villages under study. This might be a factor in farmers gathering information from sources available in their close proximity such as input retailers and company agents, etc. instead of going to these public-sector offices. Most of the farmers expressed that these officers seldom or never visited their village. However, few farmers pro-actively contacted and sought information through phone on various farm related issues from KVK-SMS on a regular basis.

Data regarding information seeking behaviour from personal cosmopolite sources is presented in Table 18. It shows that the majority of respondents (57.73%) showed high level of information seeking behaviour from personal cosmopolite sources. About 35.77 per cent of respondents were categorized under medium information seeking behaviour category from personal cosmopolite sources. Only 6.50 per cent of respondents showed low information seeking behaviour from personal cosmopolite sources.

Table 18: Distribution of respondents on the basis of information seeking behaviour from personal cosmopolite sources (n=246)

S. No.	Personal cosmopolite	Frequency	Percentage
1.	Low (Less than 10)	16	6.50
2.	Medium (In between 10 and 12)	88	35.77
3.	High (More than 12)	142	57.73

The present findings stand contradictory to those of **Raghuvanshi (2015)** who reported that exactly 50 per cent of the respondents showed medium level of information seeking behaviour from personal cosmopolite sources followed by about 41.82 per cent and only 8.18 per cent of respondents who had low and high levels of information seeking behaviour from personal cosmopolite sources respectively. **Verma and Sharma (2015), Basera (2019)** and **Dash (2019)** recorded similar contradictory observations where most of the farmer respondents showed medium level of information seeking behaviour from personal cosmopolite sources.

4.1.8.3. Information seeking behaviour from mass media sources

The data in Table 19 represent the distribution of respondents according to their frequency of seeking information from mass media sources. The data indicates that mobile phones, television and newspaper were the top three sources used for seeking agricultural information on regular basis by most of the respondents (45.12%, 41.06% and 21.54% respectively). While on the basis of Weighted Mean Scores (WMS), television was rated as the topmost mass medium source used by the farmer respondents with the score of 2.26. According to the WMS, the rest of the sources were found to be in the order as mobile phones, newspaper, radio, natak/plays and farm magazines with the scores of 2.24, 1.82, 1.39, 1.34 and 1.21 respectively. Informal discussions revealed that almost all the farmers had access to television from which they obtained farm information. Some of them referred to regional news channels, DD Kisan and DD Punjabi channels particularly for agricultural updates and information. Television was followed by mobile phone in terms of most widely used information source. The farmers utilized SMS service, social media platforms on mobile phone for obtaining and sharing agricultural information. The third most widely utilized mass medium was newspaper, one of the cheapest and most easily accessible information sources across the farm families in the study area. The newspapers published in the regional language were popular among farmers for seeking agriculture related information.

Table 19: Distribution of respondents according to frequency of seeking information from mass media sources (n=246)

S. No.	Mass media sources	Seeking behaviour						Weighted mean score (WMS)	Rank
		Regular		Occasionally		Never			
		f	%	f	%	f	%		
1	Newspaper	53	21.54	96	39.02	97	39.43	1.82	III
2	Farm magazine	0	0	53	21.54	193	78.46	1.21	VI
3	Natak/Plays	0	0	85	34.55	161	65.45	1.34	V
4	Radio	2	0.81	92	37.40	152	61.79	1.39	IV
5	Television	101	41.06	105	42.68	40	16.26	2.26	I
6	Mobile Phone	111	45.12	85	34.55	50	20.33	2.24	II
(i)	WhatsApp	80	32.52	120	48.78	46	18.70	2.14	
(ii)	Facebook	45	18.30	130	52.84	71	28.86	1.89	
(iii)	Twitter	18	7.32	49	19.92	179	72.76	1.34	
(iv)	YouTube	40	16.26	109	44.31	97	39.43	1.77	
(v)	SMS	98	39.83	60	24.40	88	35.77	2.04	

***Multiple responses were allowed**

Radio was the next most preferred mass medium for agricultural information. Only 37.40 per cent of the respondents used radio and that too occasionally because most of the farmers have switched from radio to television and mobile phones over a period of time. Natak/plays were organized by government agencies and PAU to sensitize farmers on various sensitive issues like stubble burning, farmer suicides etc. These plays were organized mainly by students from colleges and schools in villages to address the crucial issues. Although just around one-third of the farmer respondents (34.55%) were found to have come across such plays occasionally but it was believed that it could be a very persuasive medium in coming times to propagate social messages. The farm magazines were the least consulted mass medium (78.46% of the respondents never utilized) for gathering agricultural information as most of the farmers did not have access to it and they were also least interested to use it. Those who had access to the farm magazines occasionally utilized it. Usually, the magazines were mostly provided by the company agents which had detailed information about the utility of their products.

An attempt was also made to map the usage of social media and SMS services (via mobile phone) by the farmer respondents and a few interesting observations were recorded. The data indicates that WhatsApp was the most commonly used social media platform among the farmer respondents with a WMS of 2.14, followed by SMS service, Facebook, YouTube and Twitter with the WMS scores of 2.04, 1.89, 1.77 and 1.34 respectively. Within mobile phone category, 39.83 per cent of the farmers were regularly using SMS services regulated by Punjab Agricultural University (PAU), private companies like Mahindra, Tata, etc. and KVKs. They got information regarding crop-specific agricultural practices, weather predictions, etc. on regular basis in regional language as well as in Hindi and English via SMS. Social media platforms like WhatsApp, Facebook, YouTube and Twitter were also used by the farmers to gather farm related information. It was also revealed that many farmers were members of groups on platforms like WhatsApp and Facebook which provided latest information and an opportunity to discuss about various queries related to agriculture. Some of the groups were farmers only while some had linkages with agricultural scientists from as well. Apart from these, farmers also utilized YouTube for getting information related to new and better techniques for agricultural operations. A few farmers accessed Twitter for getting updates related to latest policies and technologies issued by various government and private agencies. Platforms like Facebook and Twitter were also used for raising issues by the young farmers. It was believed that social media has provided power to farmers to raise their voices at the level where they could be heard.

The results of the present study are in line with the findings of **Ansari and Sunetha (2014)** who reported that 97 per cent of farmer respondents used television and all of them (100 per cent) had access to mobile phones. **Raghuvanshi (2015)** also observed in a study that the most regularly used mass medium was television by 93.65 per cent of the respondents. Similar observation was made by **Papnai (2011)** and **Ansari *et al.* (2018)** who reported that mobile phones ranked first in terms of utilization as information source followed by television and radio respectively.

The data regarding information seeking behaviour from mass media sources is presented in Table 20. The data shows that more than three-fourth of the respondents (75.20%) belonged to medium category followed by 21.95 per cent of respondents who belonged to high category of information seeking behaviour from mass media sources. Just 2.85 per cent of respondents belonged to low category of information seeking behaviour from mass media sources.

Table 20: Distribution of respondents on the basis of information seeking behaviour from mass media (n=246)

S. No.	Mass media	Frequency	Percentage
1.	Low (Less than 14)	07	2.85
2.	Medium (In between 14 and 18)	185	75.20
3.	High (More than 18)	54	21.95

The findings of the study are supported by **Kaur (2011)** who inferred that most of the respondents had medium level of exposure to mass media as information source i.e. 76 per cent followed by high (15.5%) and low (8.5%) categories respectively. Similar findings were reported by **Devi and Verma (2011), Raghuvanshi (2015)** and **Basera (2019).**

4.1.8.4. Information seeking behaviour from extension education methods

The distribution of respondents according to their frequency of seeking information from extension education methods is given in Table 21. It can be inferred from the data that for respondents in the study area, most frequently utilized information extension method was *Kisan Mela* with weighted mean score (WMS) of 1.97. The other extension education methods followed in the order were meetings, group discussions, demonstrations, field trials and workshops with WMS of 1.55, 1.50, 1.45, 1.42 and 1.41 respectively. The reason behind these observations might be explained by the fact that the farmers were not interested in getting engaged in various extension activities due to lack of time.

Table 21: Distribution of respondents according to their frequency of seeking information from extension education methods (n=246)

S. No.	Extension education methods	Seeking behaviour						Weighted mean score (WMS)	Rank
		Regular		Occasionally		Never			
		f	%	f	%	f	%		
1	Meetings	33	13.41	71	28.87	142	57.72	1.55	II
2	Group Discussion	18	7.32	88	35.77	140	56.91	1.50	III
3	Demonstrations	11	4.47	89	36.18	146	59.35	1.45	IV
4	Field Trials	9	3.66	86	34.96	151	61.38	1.42	V
5	*Kisan Mela*	60	24.39	120	48.78	66	26.83	1.97	I
6	Workshops	5	2.03	91	37.00	150	60.97	1.41	VI

***Multiple responses were allowed**

Apart from this, some farmers expressed their inconvenience to attend such extension activities because the agencies organizing these activities were distant from their respective villages. On the other hand, the *Kisan Melas* showcased new technologies including new varieties, farm implements, etc. for which farmers were keen to know about. The farmers also attended the discussion sessions with the scientists and experts during the fairs. *Kisan Melas* offered more options for fun, frolic and recreation for all the family members. Thus, some farmers used to attend the *Kisan Melas* with their families even if they had to travel some distance for it.

The distribution of respondents on the basis of information seeking behaviour from extension education methods is presented in Table 22. It indicates that slightly less than half of respondents i.e. 47.97 per cent belonged to medium category followed by 39.02 per cent of respondents who belonged to low category. Only 13.01 per cent of respondents were found to be in high category of information seeking behaviour from extension methods.

Table 22: Distribution of respondents on the basis of their information seeking behaviour from extension education methods (n=246)

S. No.	Extension education methods	Frequency	Percentage
1.	Low (Less than 7)	96	39.02
2.	Medium (In between 7 and 11)	118	47.97
3.	High (More than 11)	32	13.01

From above data it can be observed that majority of the respondents in study area have medium and lower dependence on extension methods for seeking information. Similar findings of **Raghuvanshi (2015)** and **Basera (2019)** supported the above results.

The overall information seeking behaviour of the respondents is presented in the Table 23. It is clear from the given data that maximum percentage of the respondents (51.63%) had medium information seeking behavior followed by 27.64 per cent of respondents who had high information seeking behavior. Only 20.73 per cent of respondents showed low information seeking behaviour.

Table 23: Distribution of respondents on the basis of their information seeking behaviour (n=246)

S. No.	Information seeking behavior	Frequency	Percentage
1.	Low (Less than 46)	51	20.73
2.	Medium (In between 46 and 52)	127	51.63
3.	High (More than 52)	68	27.64

Similar findings were reported by researches conducted by **Venkatesan (2000)**, **Verma (2015)**, **Joshi (2016)**, **Mandal (2018)**, **Palvi *et al.* (2018)** and **Saklani (2018)** that most of the farmer respondents showed medium level of overall information seeking behavior.

It can be deduced from the Weighted Mean Scores across all the type of information sources that the farmers in the study area were mostly dependent on the input retail shop owners (2.30), followed by television (2.26), mobile phones (2.24), relatives (2.06) and progressive farmers (2.04) for agricultural information. Within the mobile phone category, the use of social media platforms like WhatsApp (2.14) and Facebook (1.89) along with the SMS service (2.04) was also quite frequently observed. The use of social media in particular was to share information through multimedia options like text, video, audio, etc., to raise opinions, discuss problems and find solutions to those problems.

Thus, these channels might prove helpful in communicating information to the farmers regarding measures against stubble burning. For example, a proven product as an alternative to stubble burning can be made available at the nearest input retail shop, advertised through television and social media and verbally propagated through progressive farmers and other preferred localite channels. In this way, the farmers may be exposed to a potential solution to stubble burning by using these sources or channels.

4.1.9. Innovativeness

The data regarding distribution of respondents on the basis of their innovativeness is given in Table 24. The data clearly indicates that almost half of the farmers (49.19%) exhibited medium level of innovativeness. Meanwhile, 28.05 per cent of the respondents possessed high level of innovativeness and 22.76 per cent of the farmers showed low level of innovativeness.

Table 24: Distribution of respondents on the basis of innovativeness (n=246)

S. No.	Innovativeness	Frequency	Percentage
1.	Low (less than 17)	56	22.76
2.	Medium (In between 17 and 23)	121	49.19
3.	High (more than 23)	69	28.05

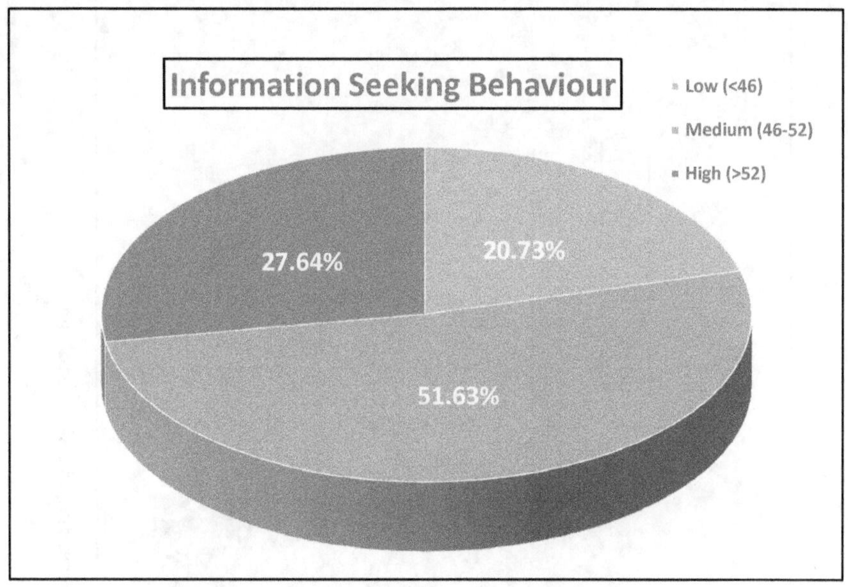

Figure 14: Distribution of respondents according to information seeking behaviour

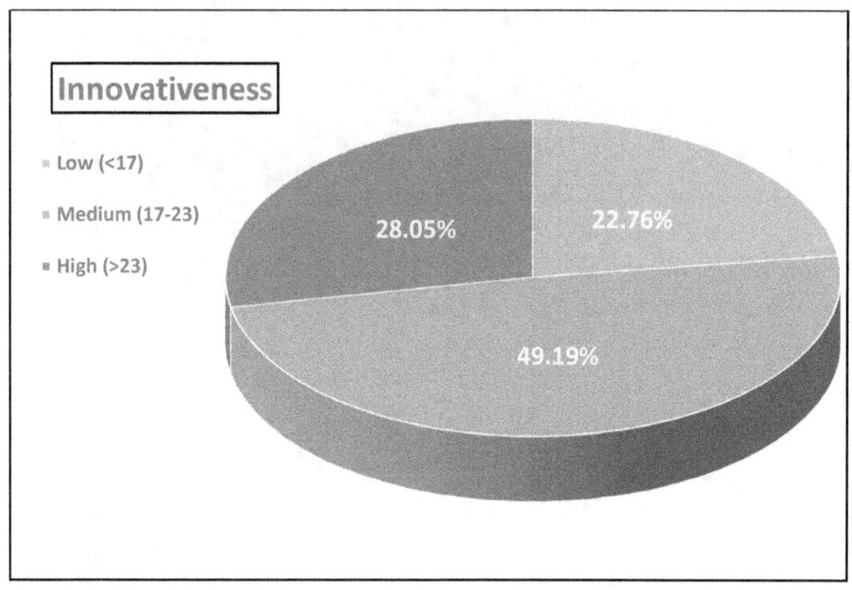

Figure 15: Distribution of respondents according to innovativeness

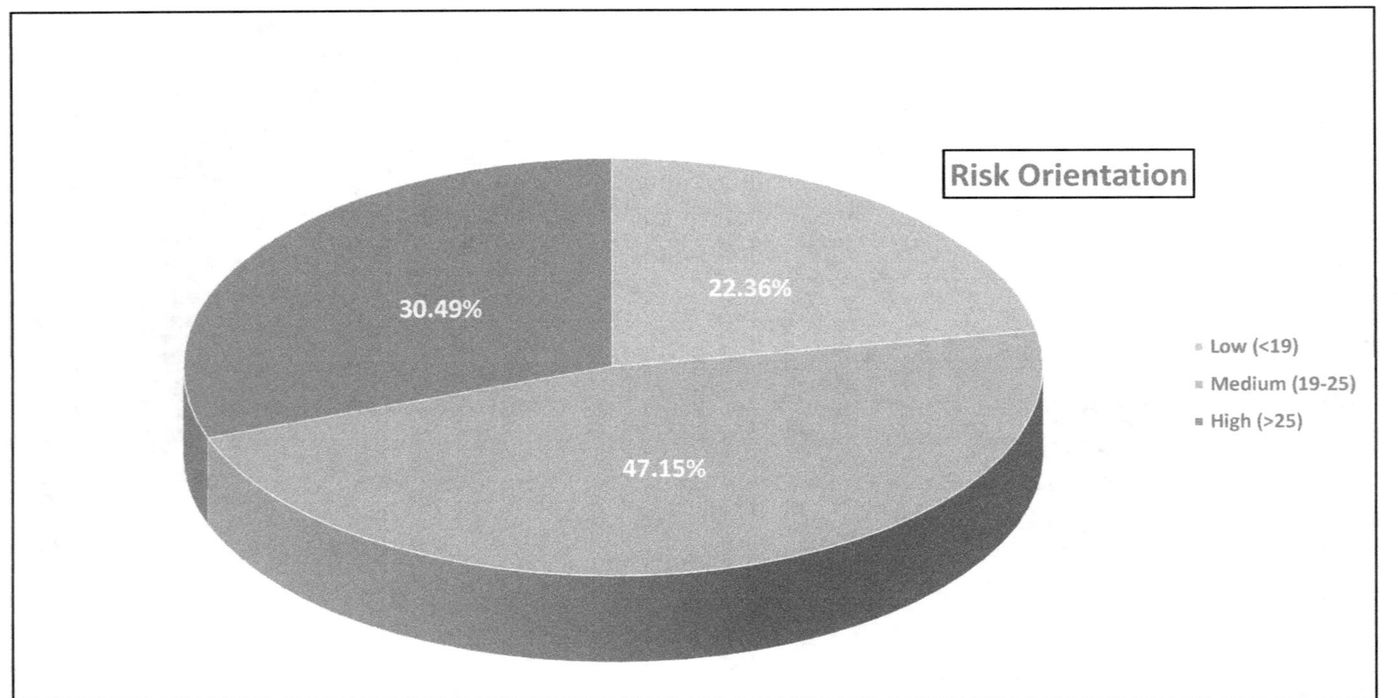

Figure 16: Distribution of respondents according to risk orientation

It can be inferred from the findings that the farmers were willing to try new and better alternatives of stubble management than to just burn the stubble. In fact, many of them have adopted and used measures like Happy Seeder, harvesting through SMS mounted combine harvesters, etc. However, the experiences of the use of these alternatives varied from farmer to farmer and some farmers reverted back to burning the stubble.

The above findings were found to be in line with those of **Roy (2015)** who also reported that majority of the farmers in Punjab (56.67%) and West Bengal (46.67%) had medium level of innovativeness when it comes to considering the new ideas related to stubble management. The study conducted by **Kumar (2008)** and **Kumar *et al.* (2013), Sharma *et al.* (2017)** conducted in Punjab also supported the above findings. However, findings of **Lyngdoh (2018)** stand contradictory to the present findings which reported that maximum number of respondents (41.67%) possessed high level of innovativeness followed by 33.33 per cent and 25 per cent of them having medium and low level of innovativeness respectively in case of adopting new ideas related to stubble management in Punjab.

4.1.10. Risk Orientation

The Table 25 depicts the data regarding the distribution of farmer respondents on the basis of their risk orientation. It can be inferred from the following data that maximum number of respondents (47.15%) possessed medium level of risk orientation. On the other hand, 30.49 per cent farmers showed high level while 22.36 per cent farmers showed low levels of risk orientation.

Table 25: Distribution of respondents on the basis of Risk Orientation (n=246)

S. No.	Risk Orientation	Frequency	Percentage
1.	Low (less than 19)	55	22.36
2.	Medium (In between 19 and 25)	116	47.15
3.	High (more than 25)	75	30.49

During the interaction with the farmers, it was revealed that farmers having larger land holdings; mixed or diversified farming or any other source of income apart from agriculture, had options to depend on those sources for sustained income. Such farmers were enthusiastic to take even more risks from these enterprises to earn additional income. Thus, they showed medium to higher risk orientation. On the other hand, the farmers following rice wheat mono cropping system got Minimum Support Price (MSP) on both these crops, thus leading to their easy marketing. Production constraints, good marketing facilities, satisfactory income from the two crops, less efforts required on the part of the farmers as compared to other cropping practices indicates the lower risk orientation among other farmers.

The findings were found to be in line with those of **Verma (2015)** in Uttar Pradesh and **Saklani (2018)** in Uttarakhand which also reported that maximum number of farmers showed medium level of risk orientation. In case of Punjab, **Roy (2015), Sharma** *et al.* **(2017)** and **Lyngdoh (2018)** also reported that most of the farmers in Punjab showed medium to higher levels of risk orientation.

4.1.11. Scientific Orientation

The data regarding distribution of respondents on the basis of their scientific orientation has been presented in Table 26. The data clearly reveals that majority of the farmers in the study area (57.32%) showed medium level of scientific orientation, followed by farmers belonging to 'high' category of scientific orientation who constituted 25.61 per cent of the total sample. It was also found that 17.07 per cent farmers exhibited 'low' level of scientific orientation.

Table 26: Distribution of respondents on the basis of Scientific Orientation (n=246)

S. No.	Scientific Orientation	Frequency	Percentage
1.	Low (less than 20)	42	17.07
2.	Medium (In between 20 and 25)	141	57.32
3.	High (more than 25)	63	25.61

Punjab is one of the most agriculturally advanced states in the country. The state has thrived upon the success of Green Revolution of 1960s and led the farmers to turn to scientific methods of agriculture. Although, the re-inventions and farm condition-specific modification of scientific techniques and equipments is the part and parcel of the farmers of Punjab but they do believe that scientific outlook has helped them to come a long way since the Green revolution in terms of increased incomes and improved standards of living. This could be clearly seen in the current findings as well, which depict that majority of the farmers showed medium to high level of scientific orientation.

The current findings stand in line with those of **Johnson and Monoharan (2007)**, **Pandey *et al.* (2011)** and **Kumar *et al.* (2013)** which also reported similar results. The study conducted by **Lyngdoh (2018)** in Punjab also reported that maximum number of farmers (40%) had a medium level of scientific orientation followed by 35 and 25 per cent of the farmers having high and low level of scientific orientation respectively.

4.1.12. Ecological Consciousness

It is evident from the data in Table 27 that maximum respondents were categorized under high category of ecological consciousness (45.53%). It was also found that more than a quarter of respondents (26.42%) showed medium level while around one-fifth of the respondents (19.92%) exhibited low level of ecological consciousness. The farmers in the study area were found to be aware of the importance of a healthy environment and that stubble burning lay harmful effects on soil and human health. But they consider other pressing issues which affected them immediately like those related with the timely sowing of the wheat crop and costly available alternatives to manage the stubble.

Table 27: Distribution of respondents on the basis of Ecological Consciousness (n=246)

S. No.	Ecological Consciousness	Frequency	Percentage
1.	Low (less than 15)	59	19.92
2.	Medium (In between 15 and 19)	75	26.42
3.	High (more than 19)	112	45.53

The studies conducted in Punjab by **Roy (2015)** and **Lyngdoh (2018)** reported similar results. They also found that majority of the farmer respondents (more than 60% in both cases) showed high levels of ecological consciousness.

4.1.13. Economic Motivation

The perusal of data in Table 28 indicates the distribution of respondents according to their economic motivation. It can be inferred from the data that maximum number of respondents (42.28%) possessed high economic motivation. Further, it was observed that 36.99 per cent of the farmer respondents were categorized under medium and 20.73 per cent respondents were placed under low economic motivation categories.

Table 28: Distribution of respondents on the basis of Economic Motivation (n=246)

S. No.	Economic Motivation	Frequency	Percentage
1.	Low (less than 19)	51	20.73
2.	Medium (In between 19 and 25)	91	36.99
3.	High (more than 25)	104	42.28

It was observed during the course of the study that the economic gains mattered a lot to the farmers in the study area. Most of the farmers were high on economic motivation as their mind set indicates that high economic status brings more respect and prestige in the society. Hence, the farmers who had lesser incomes were working hard to increase it while who were better off, were trying to at least maintain or further elevate their status. A very few farmers exhibited the virtues of satisfaction and contentment with whatever they had.

The findings reported in the works of **Joshi (2016)** and **Lyngdoh (2018)** stand in line with the present findings as both of them found that most of the farmers showed high economic motivation in Uttar Pradesh and Punjab respectively. On the other hand, **Verma (2015), Palvi *et al.* (2018)** and **Saklani (2018)** found that the farmer respondents in their respective studies in Uttar Pradesh, Madhya Pradesh and Uttarakhand exhibited medium levels of economic motivation.

4.1.14. Awareness related to stubble management measures

The data in Table 29 depicts the distribution of respondents on the basis of their awareness regarding various stubble management measures. It is evident from

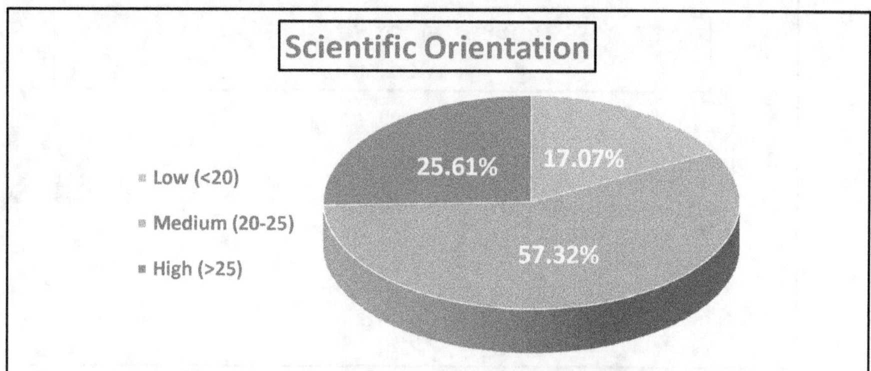

Figure 17: Distribution of respondents according to scientific motivation

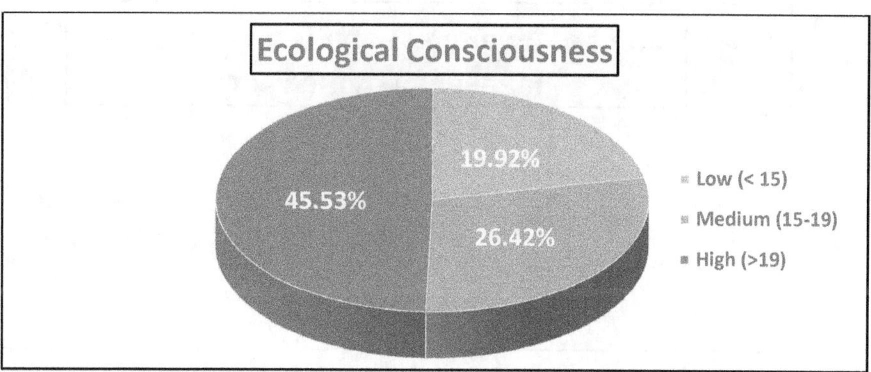

Figure 18: Distribution of respondents according to ecological consciousness

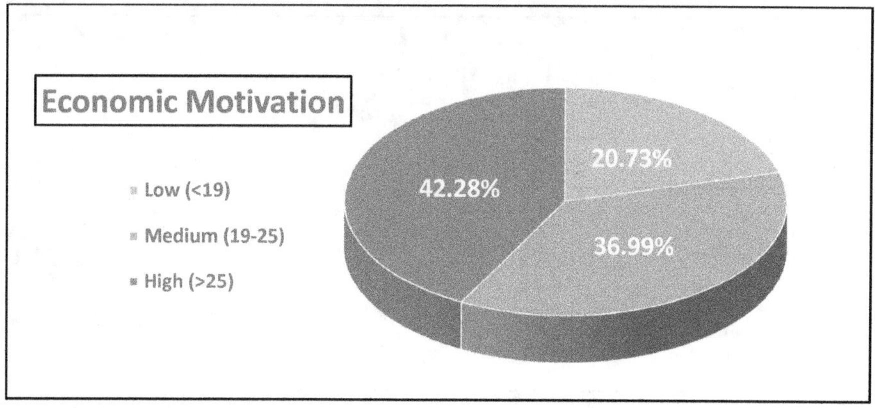

Figure 19: Distribution of respondents according to economic motivation

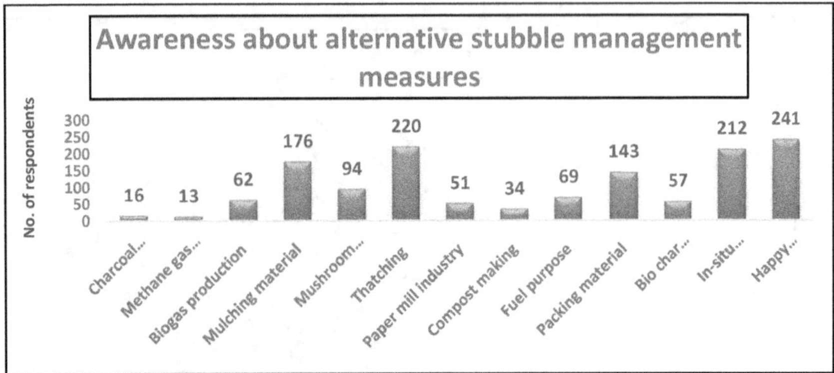

Figure 20: Distribution of respondents according to awareness about alternative stubble management measures

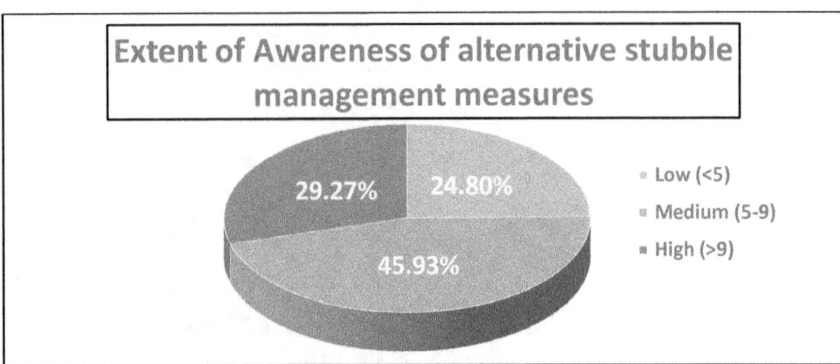

Figure 21: Distribution of respondents according to extent of awareness about alternative stubble management measures

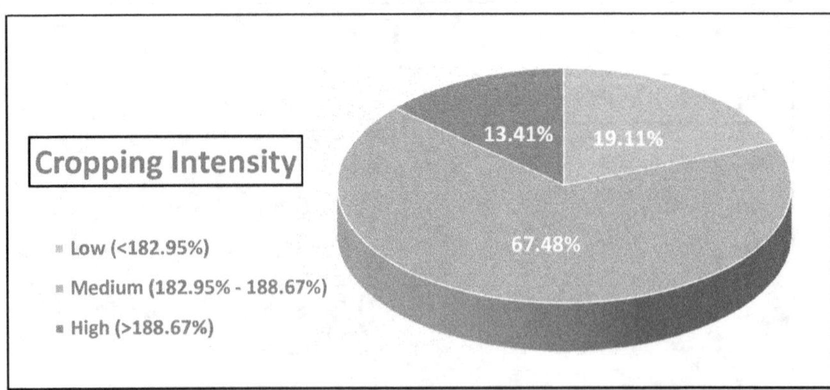

Figure 22: Distribution of respondents according to cropping intensity

the following data that majority of respondents (97.96%) were aware about Happy Seeder/Zero tillage, followed by thatching (89.43%), in-situ incorporation (86.17%) and use of stubble as mulching material (71.54%). The rest of the measures in the order of decreasing level of awareness were found as: use of stubble in packing material (58.13%), mushroom production (38.21%), fuel purposes (28.04%), biogas production (25.20%) and bio char production (23.17%). These alternatives were followed by other stubble management measures like compost making (13.82%), charcoal production (6.50%) and methane gas production (5.28%).

Table 29: Distribution of respondents on the basis of awareness related to stubble management measures (n=246)

S. No.	Stubble Management Measures	Frequency	Percentage
1	Charcoal production	16	6.50
2	Methane gas production	13	5.28
3	Biogas production	62	25.20
4	Mulching material	176	71.54
5	Mushroom production	94	38.21
6	Thatching	220	89.43
7	Paper mill industry	51	20.73
8	Compost making	34	13.82
9	Fuel purpose	69	28.04
10	Packing material	143	58.13
11	Bio char production	57	23.17
12	In-situ incorporation	212	86.17
13	Happy seeder/Zero tillage	241	97.96

***Multiple responses were allowed**

4.1.14.1. Extent of awareness related to stubble management measures

Further, the data in Table 30 depicts the extent of awareness of respondents about various stubble management measures. It can be observed from the data that maximum number of respondents (45.93%) was having medium level of awareness, i.e. out of above mentioned 13 stubble management measures, they were aware of five to nine measures. On the other hand, about 29.27 per cent of the respondents were aware of more than nine stubble management measures and thus, belonged to high awareness level. It was also observed that just less than a quarter of respondents (24.80%) were aware of less than five stubble management measures. Hence, they were classified under low awareness category.

Table 30: Distribution of respondents on the basis of extent of awareness related to stubble management measures (n=246)

S. No.	Extent of Awareness about stubble management measures	Frequency	Percentage
1.	Low (less than 5)	61	24.80
2.	Medium (In between 5 and 9)	113	45.93
3.	High (more than 9)	72	29.27

The farmers expressed that although they had awareness about various stubble management measures but they were not using them as none of them was cost efficient and more convenient than burning. This was also stated in the study conducted by **Roy and Kaur (2016)** in Punjab.

4.1.15. Cropping Intensity

The data regarding distribution of respondents according to their cropping intensity has been presented in Table 31. The data reflects the majority of the farmers (67.48%) were having cropping intensity between 182.95% to 188.67% and categorized under medium category. The farmers having low cropping intensity, i.e. below 182.95% constituted 19.11 per cent of the total sample. On the other hand, 13.41 per cent of the farmers had high cropping intensity, i.e. more than 188.67%.

Table 31: Distribution of respondents on the basis of Cropping Intensity (n=246)

S. No.	Cropping Intensity	Frequency	Percentage
1.	Low (less than 182.95)	47	19.11
2.	Medium (In between 182.95 and 188.67)	166	67.48
3.	High (more than 188.67)	33	13.41

Cropping intensity indicates, on average, the number of times the cultivated area was sown during the year. A higher cropping intensity indicates that the cropping area is being sown more than once during a year. The average cropping intensity in the state of Punjab is about 189% **(Economic Survey of Punjab, 2019-20)**. Hence, the present findings are in the similar lines with the state figures.

Hence, the above section revealed the profile characteristics of the farmer respondents in the study area. The profile characteristics included socio-personal, economic, communication, psychological and situational characteristics of the farmers. The socio-personal characteristics of the respondents can be summed up as: most of the respondents were middle-aged (65.46%) between 26-50 years, male (91.06%) and had formal education up to intermediate level (21.95%). The results related to economic characteristics revealed that majority of the farmers (51.46%) belonged to low annual income category i.e. between Rs. 50,000 to Rs. 4,50,000 per annum, possessed semi-medium land holding (27.24%), i.e. between 5-10 acres, lager milch animal like cows and buffaloes (98.78%) and were practising specialized farming (45.93%).

The information seeking behaviour of the respondents was assessed under four channels: personal localite, personal cosmopolite, mass media and extension education methods. The findings showed that relatives, progressive farmers and neighbours were the most frequently contacted personal localite sources of agricultural information. Majority of the respondents (70.73%) were categorized under medium level of information seeking behaviour from personal localite sources. Further, it was revealed that among personal cosmopolite channels, the input retail dealers/shopkeepers and company agents were most frequently contacted by the farmers of the study area. Majority of the farmers (57.73%) showed high level of information seeking behaviour from personal cosmopolite channels. The television, mobile phones and newspaper were the top-rated mass media for seeking agricultural information. The farmers were also found to use SMS services, social media platforms, e.g. WhatsApp, Facebook and YouTube etc. for sharing information among each other. A high majority of the farmer respondents (75.70%) fell under medium level of information seeking behaviour from mass media sources. Lastly, it was also found that *Kisan Melas* and meetings were the most frequent extension education methods utilized by the farmers in the study area. A large number of farmers (47.97%) exhibited medium level of information seeking behaviour from extension education methods. Overall, 51.63 per cent of the farmers were categorized under medium information seeking behaviour.

The findings also revealed that under psychological and situational characteristics, most of the farmers possessed medium level of innovativeness (49.19%), risk orientation (47.15%) and scientific orientation (57.32%) while high levels of ecological consciousness (45.53%) and economic motivation (42.28%). It was also revealed that most of the farmer respondents (45.93%) showed medium level of awareness towards various stubble management measures and had medium cropping intensity (67.48%).

These profile characteristics were crucial in designing appropriate social marketing plan as an intervention in the forthcoming objectives.

4.2. STUBBLE BURNING BEHAVIOUR OF THE FARMERS

The present study featured the application of Theory of Planned Behaviour (TPB) in order to explain the Stubble burning behaviour of the farmers. According to the TPB, the behaviour under concern is determined by the intention to perform the behaviour, which in turn is governed by various factors like attitude, subjective norms and perceived behavioural control. Therefore, this section describes the findings related to TPB constructs taken for the study. All the TPB constructs namely attitude towards stubble burning, subjective norms associated with stubble burning, perceived behavioural control related to stubble burning, behavioural intention regarding stubble burning and lastly, the stubble burning behaviour have been described below.

4.2.1. Attitude towards Stubble Burning

Attitude referred to the overall evaluation of positive or negative aspects of stubble burning by the farmer. The data in Table 32 depicts the distribution of the farmer respondents based on their attitude towards stubble burning. The perusal of the data indicates that majority of the respondents (51.63%) showed positive attitude towards stubble burning, followed by neutral (35.37%) attitude and negative (13.00%) attitude towards stubble burning.

Table 32: Distribution of respondents on the basis of Attitude towards Stubble Burning (n=246)

S. No.	Attitude towards Stubble Burning	Frequency	Percentage
1.	Negative (less than 22)	32	13.00
2.	Neutral (In between 22 and 26)	87	35.37
3.	Positive (more than 26)	127	51.63

The majority of the respondents were found to be in the favour of stubble burning as they considered burning as the most economical of all the available stubble management measures. Further, burning the stubble also helps in timely sowing of the subsequent wheat crop after paddy harvesting. Most of the farmers believed that burning the stubble kills some of the weeds and pests which, otherwise, emerge again in succeeding crops after taking shelter in the left-over stubble on the fields. So, they considered it wise to burn the paddy stubble after harvesting.

On the other hand, some farmers with negative attitude were completely against the practice of stubble burning. According to such farmers, the stubble burning should be banned. Some of the respondents also believed that stubble burning causes considerable environmental pollution. According to some farmers, stubble burning affected the farm families the most. The breathing problems, severe irritation in eyes, asthma etc. were the common health issues encountered by the farm families during the burning season, especially by the old, sick and children.

However, as the majority of farmers looked at the positives and the benefits of the stubble burning, while others were concerned with the ill-effects of burning on human health and environment. Some of the respondents agreed to both, positive and negative aspects of the stubble burning and thus, were classified under 'neutral' attitude category. The findings were found to be in line with those of **Coffie (2015)** which also reported positive attitude of the respondents towards the behaviour in their study which involved the use of Theory of Planned Behaviour.

4.2.2. Subjective norms associated with Stubble Burning

Subjective norms referred to the overall compliance with the influence of the social pressure of significant people on the respondents, regarding stubble burning. The data in Table 33 shows the distribution of respondents according the strength of subjective norms associated with Stubble Burning. The data reveals that the maximum number of respondents (42.68%) possessed strong level of subjective norms associated with stubble burning. On the other hand, 41.47 percent of farmer respondents had moderate level whereas 15.85 per cent of them had low level of subjective norms associated with stubble burning.

Table 33: Distribution of respondents on the basis of Subjective norms associated with Stubble Burning (n=246)

S. No.	Subjective norms associated with Stubble Burning	Frequency	Percentage
1.	Weak (less than 10)	39	15.85
2.	Moderate (In between 10 and 14)	102	41.47
3.	Strong (more than 14)	105	42.68

The results may be explained by the fact that most of the farmers imitated the actions of fellow farmers. During the course of investigation, it was revealed that amid restrictions on stubble burning, it was basically a group imitation and farmers followed each other if they had to burn the stubble. So, mostly the farmers relied on their fellow farmers or neighbouring farmers for this particular practice as there was a chance of incident being reported to the police and other authorities. If one of the farmers burnt the stubble, then the others would also do the same. Since most of the farmers were in favour of the idea of stubble burning, this affected the overall practice in the study area.

It was also reported that the farmer groups and unions in study area voiced in support of the burning as they were not able to bear the benefits of the schemes/solutions provided by the government to avail other machine-based measures. Some of the farmers argued that they failed to receive full benefits of subsidies provided by the government to the farmer cooperatives/groups for purchasing the set of machineries for managing the stubble as majority of the monetary amount was taken up by the middlemen.

Farmers also expressed that some of them have already tried alternative measures like the use of Happy Seeder etc. However, it didn't meet their expectations as it allowed weed infestation in subsequent crop and reduced germination was noticed for the next crop.

Another bunch of farmers were contacted by private mill owners to produce square shaped bales of the stubble which were supposed to be purchased by them at a fixed rate. The machine to covert stubble into bales was provided by the mill owners on rent basis which was paid by the farmers at the rate of Rs. 1000-2000 per acre. But after making the bales, the mill owners didn't collect all of the bales from the farmers' fields. This increased the overall expenses and added economic burden on the farmers which didn't bear any fruitful consequences. Thus, they were bound to burn those left-over bales.

On the other hand, some of the well-off farmers in the study area who didn't support the idea of stubble burning have turned to another machine called Super Seeder for managing the stubble. The basic difference between Happy Seeder and Super Seeder is that the former chops the stubble and spreads it evenly on the ground while the latter chops and churns the stubble into the soil, thus, giving satisfactory results. Interestingly, there was a difference of opinion among the users of Super Seeder as well. While some were satisfied by its overall function and performance, some argued that the incorporation of the stubble (in general by any method or machine like rotavator etc.) results in yield reduction of the subsequent crop. Also, there was no subsidy on Super Seeder at the time of investigation.

However, the popular opinion and the existing subjective norms among the farmer respondents was in the favour of stubble burning as many farmers have reverted back to stubble burning after trying various alternatives. Thus, farmers' decision regarding stubble burning was widely affected by their peers and most of them favoured the practice of stubble burning.

The present findings were found to be in contradiction with **Coffie (2015)** which reported moderate level of subjective norms associated with the behaviour under concern.

4.2.3. Perceived Behavioural Control related to Stubble Burning

Perceived behavioural control referred to the perceived ease of burning the stubble. The data related to Perceived Behavioural Control regarding Stubble Burning is presented in Table 34. It shows that the majority of the respondents (58.54%) showed 'high' degree of Perceived Behavioural Control, followed by farmers showing 'medium' (27.23%) and 'low' (14.23%) degree of Perceived Behavioural Control respectively.

Table 34: Distribution of respondents on the basis of Perceived Behavioural Control related to Stubble Burning (n=246)

S. No.	Perceived Behavioural Control related to Stubble Burning	Frequency	Percentage
1.	Low (less than 16)	35	14.23
2.	Medium (In between 16 and 20)	67	27.23
3.	High (more than 20)	144	58.54

The results might be explained by the fact that burning the stubble is practically the easiest measure to manage it after harvesting, thus, giving high degree of perceived behavioural control to the farmer respondents.

During the time of investigation, it was found that there were official restrictions on stubble burning in the states of Punjab and Haryana. Farmers burning the paddy stubble after harvesting were liable to be booked under IPC Section 188. The punishment involved a monetary fine which may extend to Rs. 200 per acre or imprisonment for up to one-month period or both. However, during the investigation, it was revealed that there was no on-spot monitoring for stubble burning incidents by the local law-enforcing bodies in the study area at that time. Amid the official restrictions, farmers felt that they could afford the monetary penalties. Thus, lack of

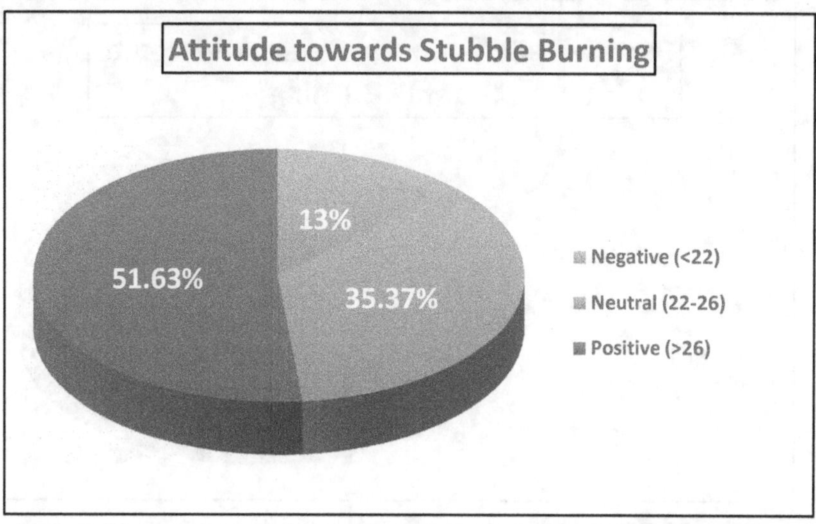

Figure 23: Distribution of respondents according to attitude towards stubble burning

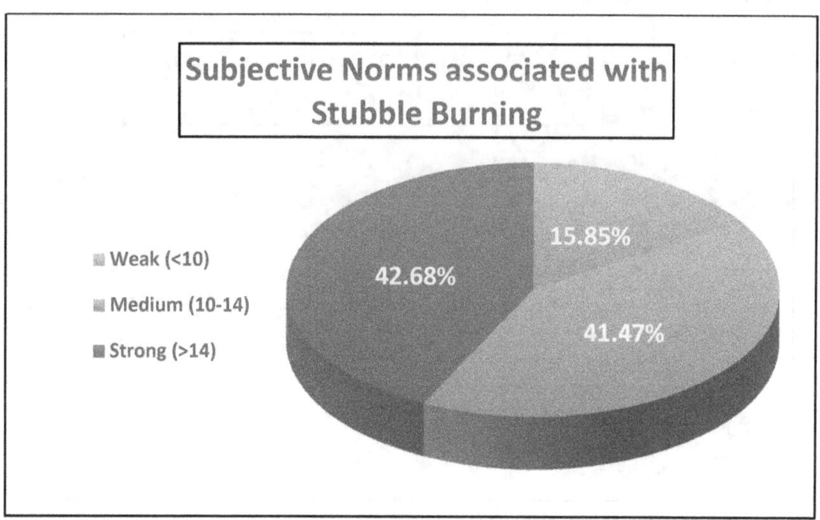

Figure 24: Distribution of respondents according to subjective norms associated with stubble burning

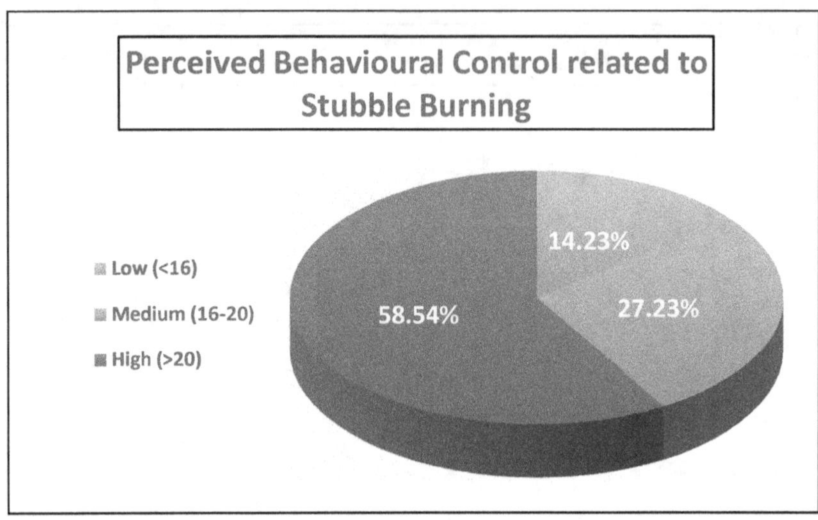

Figure 25: Distribution of respondents according to perceived behavioural control related to stubble burning

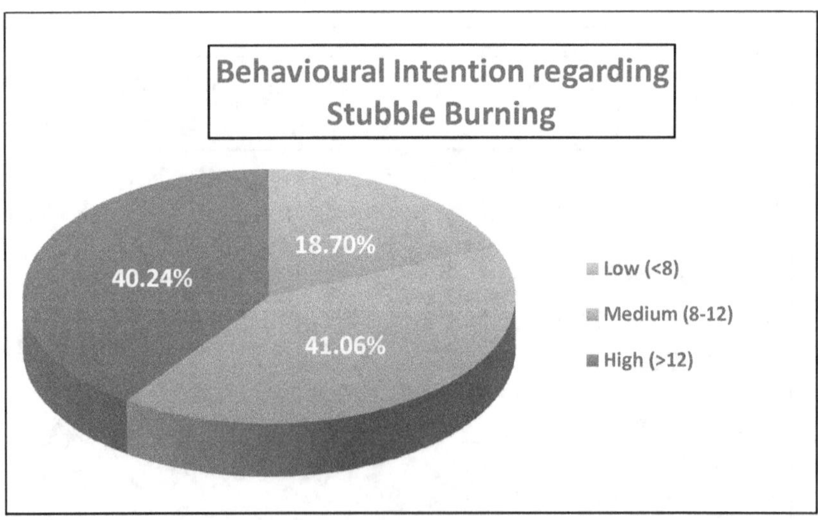

Figure 26: Distribution of respondents according to behavioural intention regarding stubble burning

monitoring and affordable penalties added to the convenience for farmers for burning the stubble, thereby providing them a higher degree of perceived behavioural control for stubble burning.

The discussions with the farmers also revealed that the procedure to apply for subsidies provided by the government on machines and implements to the farmer societies was very cumbersome and was not fruitful. According to the schemes, a subsidy of 80 per cent was to be provided to the registered farmer societies on the purchase of a combo-set of farm implements including Happy Seeder, Mulcher, Plough and Chopper. The overall cost of the combo-set was estimated around Rs. 25 lakhs. After the full payment at the time of purchase, the subsidies were to be directly transferred to the bank accounts of the farmers. However, most of the farmers had little trust on the procedure and felt that their money will be curbed in the process by the middlemen. They cited examples of other farmer societies from nearby villages which didn't receive any subsidies even after a year of application. All this has led the farmers to stick to the easy method of stubble burning in the study area for managing the paddy stubble.

The findings of present study were found to be in contradiction with those of **Coffie (2015)** which reported low degree of Perceived Behavioural Control related with the behaviour under concern.

4.2.4. Behavioural Intention regarding Stubble Burning

Behavioural intention referred to the willingness or readiness of the farmers to burn the paddy stubble after harvesting. The data in Table 35 indicates the distribution of farmer respondents on the basis of their Behavioural Intention regarding Stubble Burning. It can be inferred from the data that maximum number of respondents (41.06%) were categorised under 'medium' level of readiness or intention to burn the stubble after harvesting at the time of investigation. It was further noted that almost equal proportion of farmers were categorized under 'high' category (40.24%). Lastly, those farmers who were categorised under 'low' category constituted only 18.70 per cent of the total respondents.

Table 35: Distribution of respondents on the basis of Behavioural Intention regarding Stubble Burning (n=246)

S. No.	Behavioural Intention regarding Stubble Burning	Frequency	Percentage
1.	Low (less than 8)	46	18.70
2.	Medium (In between 8 and 12)	101	41.06
3.	High (more than 12)	99	40.24

During the period of investigation, it was found that most of the farmers were ready to burn the stubble in that on-going season, despite of penalties and restrictions. Hence, they were categorized under 'high' category. Many respondents also expressed that they will probably continue burning the stubble in the foreseeable future as well, amid the dearth of viable and affordable alternative stubble management measures.

However, some of the farmer respondents were ready to give up stubble burning if they were provided with suitable and better alternative measures. Such farmers were categorised under 'medium' category. Apart from these, most of the farmers who were categorised under 'low' category were not ready to burn the stubble as they were well off. Thus, they could afford other mechanical measures and were also concerned about the environment and other harmful effects of stubble burning brings to the human health. The above findings were found to be in line with those of **Coffie (2015)** which reported high level of behavioural intention under concern.

4.2.5. Stubble Burning Behaviour

The Stubble Burning Behaviour of the farmers for the study was defined according to the norms of Theory of Planned Behaviour (Ajzen, 2019) as the "act of burning the paddy stubble on the field after harvesting and before sowing of the subsequent wheat crop." In context of present study, the stubble burning behaviour was studied using two components, i.e. frequency of engagement in stubble burning behaviour and the proportion of the total produced paddy stubble managed through burning. The 'frequency' referred to the number of occasions on which the farmer burnt the stubble on field after harvesting in past five years (considering one occasion per year as paddy is grown once a year in the study area). While the 'proportion' referred to the part of the total produced stubble which was managed through burning, in case other stubble management methods were also used by the farmer.

The findings regarding stubble burning behaviour of the farmers have been described in the following sub-sections:

4.2.5.1. Frequency of engagement in Stubble Burning Behaviour

The data in Table 36 indicates the distribution of respondents based on frequency of engagement in stubble burning behaviour. The perusal of the data states that majority of farmer respondents were engaged in stubble burning every year in the last five years, thus classified under 'very high' frequency (58.13%). Around a quarter of farmer respondents (24.80%) burnt the paddy stubble after harvesting in four out of five years while 9.76 per cent of the farmer respondents had burnt the stubble in last three out of five years, thus categorised under 'high' and 'medium' frequency categories respectively. The data also shows that 4.06 per cent of the farmer respondents burnt the stubble on field twice in last five years while 3.25 per cent of the respondents had burnt the stubble just once during the time period which were put under 'low' and 'very low' frequency categories respectively.

Table 36: Distribution of respondents on the basis of frequency of engagement in Stubble Burning Behaviour (n=246)

S. No.	Frequency of engagement in Stubble Burning Behaviour since past 5 years	Frequency	Percentage
1.	Very low (Once)	8	3.25
2.	Low (Twice)	10	4.06
3.	Medium (Thrice)	24	9.76
4.	High (Four times)	61	24.80
5.	Very High (Five times)	143	58.13

The farmers who did not burn the stubble, tried alternative stubble management measures like zero tillage, in-situ incorporation, Happy Seeder, Super Seeder, making bales etc. However, they reverted back to burning the stubble as the results of the alternative measures were not satisfactory to them. Either it added excessively to the cost of cultivation or lead to more weed and pest infestation in the succeeding crops. On the other hand, most of the farmers categorised under 'very low', 'low' and 'medium' frequencies of stubble burning, were those who have quitted and had not burnt the stubble in the past four, three and two years respectively. They switched to alternative measures such as use of Super Seeder etc. since then. The discussions with the farmers revealed that they have quitted burning because of the harmful effects it lays on human health and the environment. It was also found that these farmers were comparatively well-off than other farmers and could afford the extra expenditure that occurred due to extra tillage operations, fuel charges etc. while using alternative mechanical measures.

4.2.5.2. Proportion of the total produced paddy stubble managed through burning

The second component of the stubble burning behaviour was the proportion of the total produced stubble managed through burning. The results of this component have been presented in the Table 37.

Table 37: Distribution of respondents on the basis of proportion of the total produced paddy stubble managed through burning (n=246)

S. No.	Proportion of the total produced stubble managed through burning	Frequency	Percentage
1.	Very little (Less than 20%)	13	5.28
2.	Little (20% to 40%)	15	6.10
3.	Medium (40% to 60%)	26	10.57
4.	Large (60% to 80%)	71	28.86
5.	Very large (More than 80%)	121	49.19

The data indicates that the maximum number of respondents (49.19%) burnt more than 80 per cent of the total stubble produced on their field and were classified

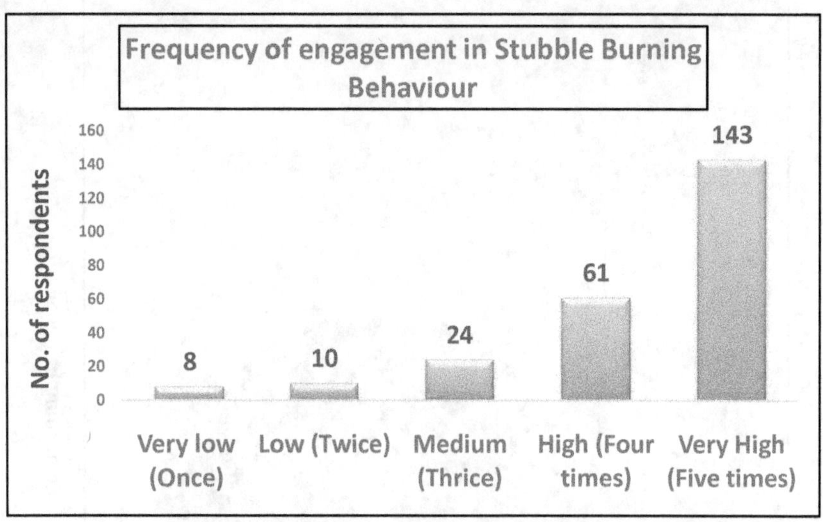

Figure 27: Distribution of respondents according to frequency of engagement in stubble burning behaviour

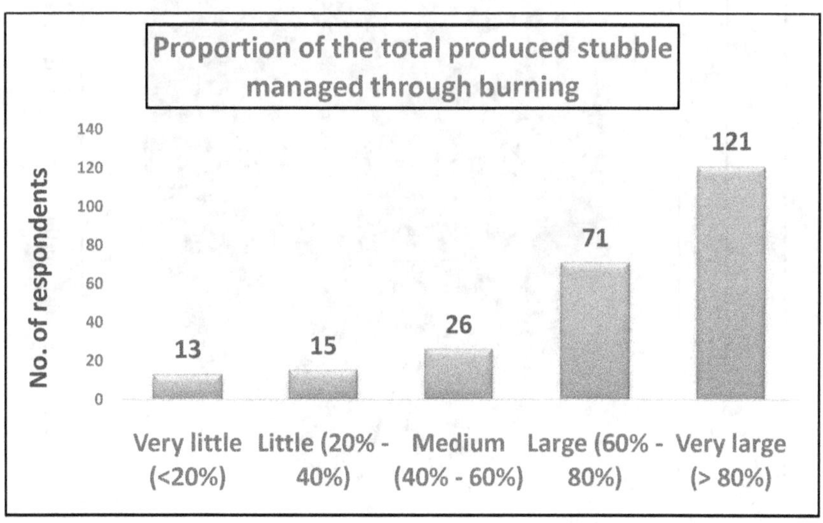

Figure 28: Distribution of respondents according to proportion of total stubble managed through burning

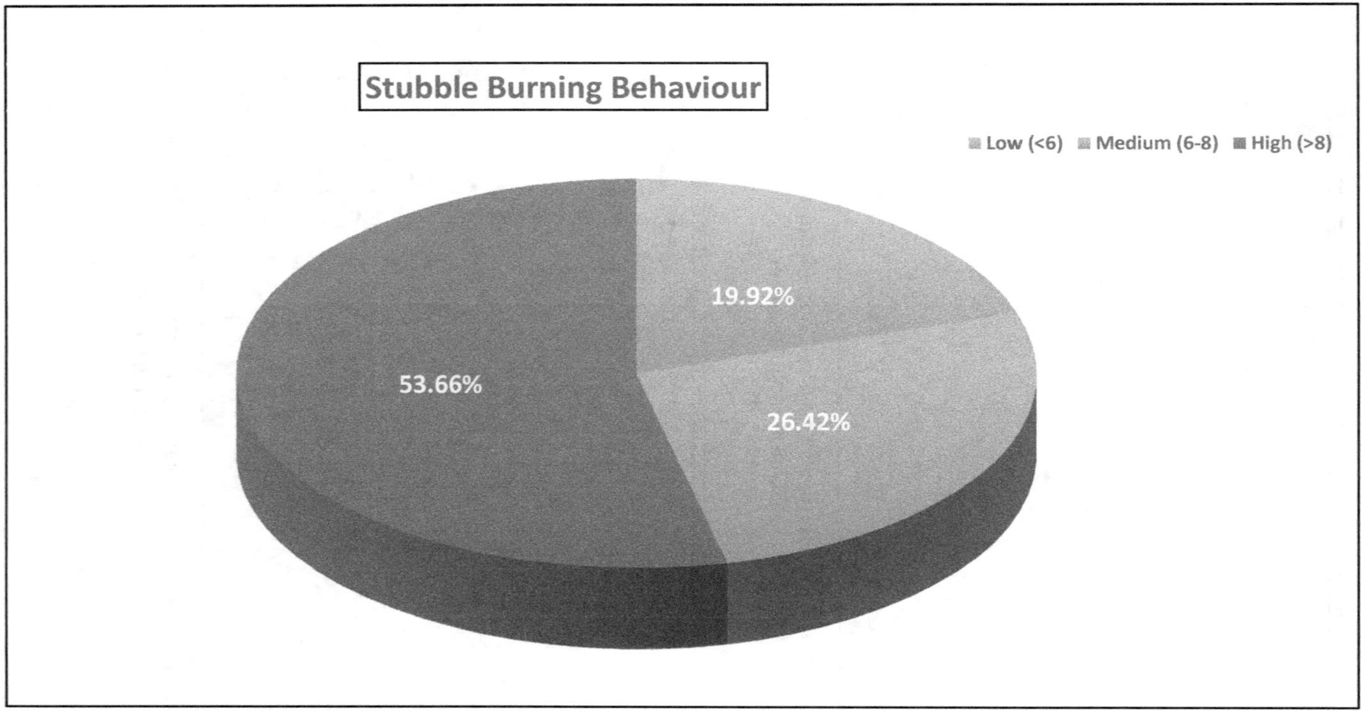

Figure 29: Distribution of respondents according to stubble burning behaviour

under 'very large' proportion category. They were followed by the farmers who burnt 60-80 percent of the total stubble produced, thus classified under 'large' proportion category. The farmers who burnt 40-60 per cent (medium), 20-40 per cent (little) and less than 20 per cent (very little) proportions of the total stubble produced on their fields accounted for 28.86 per cent, 10.57 per cent, 6.10 per cent and 5.28 per cent respectively.

During the course of investigation, it was revealed that there were various reasons for the above distribution of the proportion of the stubble burnt by the farmers. The biggest proportion of farmers preferred burning the paddy stubble after harvesting irrespective of situations. On the other hand, some of the farmers had transplanted paddy on different fields on different dates which resulted in the later maturity of the crop which was grown later. The difference between the maturity periods varied from 7-10 days to over 15 days. This gives ample time for field preparation in the early matured paddy fields but lesser time in later matured paddy fields for the sowing of subsequent wheat crop. So, some of the farmers who were well-off; managed the stubble of early matured paddy fields by alternative mechanical measures such as using Super Seeder etc. and burnt the stubble in the other later matured paddy fields.

Another reason revealed during informal discussions was that some of the farmers in Bathinda district were contacted by private mill owners for producing square-shaped bales of the stubble. So, some of the farmers tried this alternative and produced bales using baler machines provided by the mill owners. However, only few of them were contacted back for the sale of bales and most of them had to burn the square-shaped bales produced on their fields.

It was also brought to notice that some of the farmers' fields were located in such a place where burning could prove harmful to the crops growing in the nearby adjacent fields of other farmers. A considerable proportion of the farmers have left burning the stubble and switched to alternative measures, thus contributing to lower proportions of stubble managed through burning.

The scores of the respondents regarding both the above components were summed up and the respondents were classified under three categories of low, medium and high performance of stubble burning behaviour. The same is depicted in the Table 38.

Table 38: Distribution of respondents on the basis of Stubble Burning Behaviour (n=246)

S. No.	Stubble Burning Behaviour	Frequency	Percentage
1.	Low (less than 6)	49	19.92
2.	Medium (In between 6 and 9)	65	26.42
3.	High (9 and above)	132	53.66

The data indicates that the majority of farmer respondents showed 'high' performance of stubble burning behaviour (53.66%) followed by 'medium' (26.42%) and 'low' (19.92%) performers of stubble burning behaviour respectively. Beyond the numbers and figures, it was found that most of the farmers were ready to give up stubble burning in case they had another economical and suitable alternative to their needs. The above findings were found to be on contrary to those of **Coffie (2015)** which reported the low performance of the behaviour under concern by the respondents.

4.2.6. Observation on patterns of Stubble Burning

During the course of the study, the investigator observed patterns of stubble burning. It was noticed that the harvesting of the paddy crop by the mechanical combined harvester leaves the trail of stubble behind it in its line of movement. The final appearance of the harvested paddy field involves the cut and left-over stubble lying on the field in lines. The interesting thing about the observation was that the left-over stubble was burnt in one of the two ways followed by the farmers as described below:

4.2.6.1. Partial burning of the left-over stubble: The first pattern that was noticed, involved the burning of stubble lying just in the trail lines left behind by the movement of combine harvester. The rest of the stubble was churned and mixed into the soil using different implements like harrows or rotavators, etc. during the field preparation for the subsequent wheat crop.

Figure 30: Partial burning of the left-over stubble on the field

It was also revealed that in this method, the standing base of the stubble that remains un-burnt gets decomposed into the soil with time, thus adding to its enrichment. The purpose of burning the trailed stubble, according to the farmers is to kill the pest commonly known as *gobh sundi* (in the local language) which is believed to take shelter on the stubble and re-emerges when conditions are favourable. However, some of the farmers also reported that tillage operations following this kind of burning are comparatively difficult as the left over un-burnt stubble creates hindrance by sticking into the implements and takes time, fuel and energy to completely churned and mixed into the soil.

4.2.6.2. Complete burning of the left-over stubble: Another pattern involved the complete burning of the left-over stubble after spreading it evenly on the field either manually or mechanically. The whole stubble on the field is burnt after the harvesting of crop and then further operations for the next crop are carried out.

In this method, the pests are destroyed and tillage operations for the subsequent wheat crop are also carried out smoothly. Some of the farmers were aware of the drawbacks of stubble burning and potential loss it caused to soil health but still farmers were doing it because it saves efforts, time and fuel in mixing the left-over stubble into the soil and is easy for carrying out further land preparation operations.

Figure 31: Complete burning of the left-over stubble on the field

Thus, the above section revealed the findings related to constructs derived from Theory of Planned Behaviour (TPB), which forms the theoretical base of the study. It was found that maximum number of respondents showed positive attitude towards stubble burning (51.63%), strong subjective norms associated with stubble burning (42.68%), high perceived behavioural control (58.54%) and medium level of behavioural intention regarding stubble burning (41.06%).

The stubble burning behaviour of the farmers was studied using two components; namely, frequency of engagement in the stubble burning behaviour and proportion of the total stubble managed through burning. Results showed that majority of the respondents (58.13%) have been burning the paddy stubble on their field since last five or more years. It was also revealed that most of the farmers (49.19%) burnt more than 80 per cent of the total stubble produced and manages the rest by other measures such as incorporation using Super Seeder etc. It was also found that the majority of respondents (53.66%) showed 'high' performance of the overall stubble behaviour. In addition to this, an interesting observation was made on the pattern of stubble burning by the farmers. Farmers either burnt the stubble partially or completely. In the partial stubble burning, only the stubble left on trail by the mechanical harvester along its line of movement is burnt. It was done for the purpose of killing the pests and the remaining stubble is mixed into the soil. On the other hand, the complete burning of the stubble after harvesting kills the pests that are supposed to be thriving on the stubble and saves efforts, fuel and time. Complete burning also facilitates smooth land preparation operations for subsequent wheat crop.

4.3. RELATIONSHIP BETWEEN THEORY OF PLANNED BEHAVIOUR CONSTRUCTS

The previous section of the chapter covered the basic evaluation of stubble burning behaviour along with other Theory of Planned Behaviour (TPB) constructs using descriptive statistics. According to the theory, the intentions are influenced by the attitude, subjective norms and perceived behavioural control while the behaviour in question is influenced by the intentions to perform it. In the same line, one of the objectives of the present study was to find the influence of TPB constructs, i.e. attitude towards stubble burning, subjective norms associated with stubble burning and perceived behavioural control related to stubble burning on the behavioural intention of the farmers regarding stubble burning and also the influence of the behavioural intention on the stubble burning behaviour of the farmers.

In order to test this theory in context of present study, Structural Equation Modeling (SEM) was used. SEM has been widely used to analyse the Theory of Planned Behaviour constructs related to farmers and agriculture related behaviours and to test the relationships among them. The SEM consists of two models, namely; the measurement model and the structural model. The measurement model depicts how the observed (measured) variables represent the constructs. On the other hand, the structural model shows how the constructs are interrelated to each other and specifies whether a relationship between the variables exists or not.

The steps suggested by Nayak (2017) were followed to carry out the procedure of SEM using Analysis of Moment Structures (AMOS) software. The procedure consisted of six basic steps including defining the individual constructs, developing the overall measurement model, testing the measurement model, assessing the measurement model fit, specifying the structural model and lastly, assessing the structural model fit and hypothesized relationships. The related findings are described step by step as follows:

4.3.1. The Measurement Model

To evaluate the measurement model, Confirmatory factor Analysis (CFA) was used to test the reliability and validity of the constructs. CFA is a multivariate statistical procedure that is used to test how well the measured variables represent the number of constructs. It is a tool which is used to confirm or reject the measurement theory or model. The CFA was performed using the AMOS software, ver. 26. Before conducting the CFA, Statistical Package for Social Sciences (SPSS) software was used to evaluate the Kaiser-Meyen-Olkin (KMO) measure of sampling adequacy and the Bartlett's test of sphericity were applied to check the suitability of the data for factor analysis. The value of KMO was 0.721 (greater than 0.6 is recommended) and the approximate χ^2 from Bartlett's test of sphericity was found to be highly significant (p=0.0008, which is <0.001) (Hair et al., 2006) indicating that the data were suitable for the CFA.

4.3.1.1.   The validity and reliability of measurement model

To evaluate the measurement model, construct validity and reliability were examined. Construct validity refers to whether an observed variable truly measures the construct as the researcher intends it to be measured. While the reliability of a construct refers to the degree to which it yields consistent scores when it is administered number of times. The Chronbach's alpha was calculated for testing the reliability while factor loadings and Average Variance Extracted (AVE) were used to check the construct validity of the measurement model. The Cronbach's alpha values, AVE and factor loadings for all the Theory of Planned Behaviour constructs are given in Table 39.

Table 39: The validity and reliability measures of the measurement model

TPB constructs		Items	Factor loading	Average Variance Extracted	Chronbach's alpha value
Attitude towards Stubble Burning	ATT1	Burning is the most economic stubble management measure	0.932	0.763	0.727
	ATT2	Stubble burning helps in timely sowing of subsequent wheat crop	0.763		
	ATT3	Stubble burning causes environmental pollution	0.851		
	ATT4	Stubble burning is bad for human health	0.781		
	ATT5	Stubble burning is helpful in controlling weeds	0.639		
	ATT6	Stubble burning should be banned	0.672		
Subjective Norms associated with Stubble Burning	SN1	My fellow farmers support the idea of stubble burning	0.632	0.598	0.803
	SN2	The farmer groups in my area favour the practice of stubble burning	0.653		
	SN3	People who are important to me think that I should burn the stubble	0.612		
	SN4	People who influence my behavior think it is preferable to burn the stubble	0.696		

Construct	Item	Statement	Loading	AVE	CR
Perceived Behavioural Control related to Stubble Burning	**PBC1**	Burning the stubble is the easiest way to manage it	0.903		
	PBC2	There is nobody to stop or monitor the stubble burning incidents in my area	0.754		
	PBC3	The monetary penalties for burning the stubble are affordable	0.816	0.654	0.765
	PBC4	It is a cumbersome procedure to apply for subsidies for machines to manage the stubble	0.608		
	PBC5	It's totally up to me whether I burn the stubble	0.727		
Behavioural Intention regarding Stubble Burning	**BI1**	I plan to burn the stubble after the forthcoming harvesting of rice crop	0.798		
	BI2	It is likely that I will continue burning stubble in foreseeable future	0.919	0.739	0.798
	BI3	I intend to burn the stubble until a better alternative is available	0.832		
Stubble Burning Behaviour	**SBB1**	I have been burning the stubble from many years	0.867		
	SBB2	I manage a limited proportion of stubble by burning	0.792	0.752	0.775

According to Nunnally and Bernstein (1994), a minimum alpha value of 0.6 for a scale is sufficient for social science research. As the Cronbach's alpha values in Table 37 were all found to be above 0.6, the constructs were, therefore, deemed to have adequate reliability. On the other hand, the factor loadings, which represent the correlation between the item and the factor, of all items ranged from 0.608 to 0.919 which exceeded the recommended level of 0.6 (Chin *et al.*, 1997). The closer the factor loading value to 1, the more influence the factor possess on the variable. The Average Variance Extracted (AVE), which reflects the overall amount of variance in the indicators account for by the latent construct, ranged from 0.598 to 0.763, exceeding the recommended level of 0.5 as suggested by Segars *et al.* (1997). Thus, the measurement model was found to be reliable and valid.

The adequate reliability and validity of the measurement model indicates that the constructs adopted from the Theory of Planned Behaviour, i.e. attitude, subjective norms, perceived behavioural control, behavioural intention and stubble burning behaviour are apt for explaining the behavior in question.

4.3.1.2. The measurement model fitness

According to the stages of Structural Equation Modeling (SEM) analysis, the overall model was developed and all the constructs were brought together. Then, the measurement model fitness was estimated by running the AMOS software. There are three types of measurement for the model fit namely; the measure of absolute model fit, the measure of incremental model fit and the measure of parsimonious model fit (Hair *et al.*, 2006). Seven conventional model-fit statistics were chosen for the present study. The best-known index of absolute fit is the Chi-square which is denoted by χ^2/degree-of-freedom (χ^2/df). The Goodness of Fit Index (GFI) and the Comparative Fit Index (CFI) were the two indices used to evaluate the overall absolute fit of the proposed model. To evaluate the incremental fit measures, the Adjusted Goodness of Fit Index (AGFI), the Incremental Fit Index (IFI), and the Normed Fit Index (NFI) were used. Finally, the Root Mean Square Error of Approximation (RMSEA) was used to evaluate the parsimonious fitness of the proposed model. The results of these test indices are given in Table 40.

Table 40: Goodness of fit test statistics for measurement model

Fit index	χ^2/df	GFI	CFI	RMSEA	AGFI	IFI	NFI
Recommended value	<3	>0.9	>0.9	<0.05	>0.9	>0.9	>0.9
Observed value	1.067	0.932	0.956	0.048	0.958	0.972	0.961
Conclusion	Accepted	Good fit	Good fit	Good fit	Good fit	Good fit	Good fit

The data in Table 40 gives the standard recommended values and the observed values of various test indices used in the study. The recommended value for χ^2/df should be less than 3. On running the tests in AMOS software, it was found that the chi-square value for the model was 223.982 and the degree-of-freedom were 210, thus, giving the required test statistic value for χ^2/df of 1.067 which is under the recommended range. For the model to be deemed as 'fit', the observed values for GFI, CFI, AGFI, IFI and NFI should be greater than 0.9. The test results indicate that all these indices, GFI (0.932), CFI (0.956), AGFI (0.958), IFI (0.972) and NFI (0.961) were all above the desired levels. Lastly, the observed value for RMSEA was found to be 0.048 which was also within the acceptable range (<0.05). Hence, the results clearly indicate that all the indices values were acceptable and thus, the measurement model could be considered as fit and acceptable. According to the norms, once the measurement model is deemed as fit and acceptable, then only the structural model could be specified for further analysis.

4.3.2. The Structural Model

Continuing the steps suggested by Nayak (2017), the structural model was specified after the measurement model was found acceptable and fit. The next step was to test the fitness of the specified structural model and then estimating the standardized parameters. The related findings are given below in the following sub-sections:

4.3.2.1. The structural model fitness

The estimation of the structural model was done by means of AMOS using Maximum Likelihood Estimate (MLE). It is the default estimate in most of the model-fitting programs. Using MLE, all the estimates can be calculated at once. Using the similar fit indices as used in the measurement model (χ^2/df, GFI, CFI, AGFI, IFI, NFI and RMSEA), the structural model was also tested for adequate fitness. The observed values for these indices are presented in Table 41.

Table 41: Goodness of fit test statistics for structural model

Fit index	χ^2/df	GFI	CFI	RMSEA	AGFI	IFI	NFI
Recommended value	<3	>0.9	>0.9	<0.05	>0.9	>0.9	>0.9
Observed value	1.129	0.953	0.948	0.043	0.946	0.981	0.953
Conclusion	Accepted	Good fit	Good fit	Good fit	Good fit	Good fit	Good fit

The observed chi-square value for the model was 259.653 and the degree-of-freedom were 230, thus, giving the required test statistic value for χ^2/df of 1.129 which is under the recommended range (<3). Hence, it was accepted. The observed values for GFI, CFI, AGFI, IFI and NFI were found to be 0.953, 0.948, 0.946, 0.981 and 0.953. These values were well above the desired level of 0.9. Lastly, the observed value for RMSEA was found to be 0.043 which was within the acceptable range (<0.05). Thus, the perusal of data indicates that all the fitness indices had the values above acceptance levels and hence, the proposed structural model found was found to fit the data.

4.3.2.2. The structural model estimation

The structural model estimation leads to the calculation of the path estimates which represents the causal relationship between the constructs. The present study aimed at finding the strength among the hypothesised relationships between different constructs derived from the Theory of Planned Behaviour. Thus, it was checked using the SEM whether there exist any relationship between attitude, subjective norms, perceived behavioural control and behavioural intention of the farmers regarding stubble burning and that of behavioural intention on stubble burning behaviour of farmers.

4.3.3. Extent of influence of attitude, subjective norms, perceived behavioural control on behavioural intention regarding stubble burning

The findings of the estimation of structural model have been presented in Table 42. The data indicates the various path descriptions, i.e. the direction of arrow points out the influence of one variable on another. The Table also features the null-hypotheses set, the standardized path estimates, p-values obtained during analysis and the result whether the null-hypotheses were accepted or not.

Table 42: Findings of Structural Model Estimation (I)

Path Description	Standardized Path Estimates	p-values	Hypotheses	Result
ATT $\longrightarrow$ BI	0.61**	0.012	H_{01}	Rejected
SN $\longrightarrow$ BI	0.28**	0.047	H_{02}	Rejected
PBC $\longrightarrow$ BI	0.47**	0.031	H_{03}	Rejected

ATT= Attitude towards Stubble Burning, **SN**= Subjective Norms associated with Stubble Burning, **PBC**= Perceived Behavioural Control related to Stubble Burning, **BI**= Behavioural Intention regarding Stubble Burning, **p< 0.05

The description of the findings is given as follows:

4.3.3.1. Path 1: Attitude to Behavioural Intention (ATT $\longrightarrow$ BI)

It can be noted from the data that the relationship between attitude of the farmers towards stubble burning (ATT) and the behavioural intention regarding stubble burning (BI) was positive and significant. The standardized path estimate obtained was 0.61. The p-value obtained was 0.012, which is less than 0.05. Thus, it was found to be significant at 5 per cent level of significance. It meant that the behavioural intention to burn the stubble increased 0.61 units for a unit increase in the attitude towards stubble burning. This indicates that the higher degree of positive attitude of the farmers towards stubble burning made way for a greater degree of intention to burn the stubble. During the investigation, it was revealed that the timely sowing of subsequent wheat crop was of utmost importance to the farmers for which, burning the left-over stubble was the most economic measure available around. Thus, it could have affected the intention of the farmers to burn the stubble after harvesting.

It can also be noted that the attitude was the strongest influential factor among the three TPB constructs (attitude, subjective norms and perceived behavioural control) with the highest standardized path estimate value of 0.61. The findings also revealed that the hypothesis which stated that there was no significant relationship between the attitude and behavioural intention regarding stubble burning, H_{01} was rejected. Similar results were found in the studies like **Artikov *et al.* (2006), Kumar (2012), Sharifzadeh *et al.* (2012), Zhang *et al.* (2015), Sukhmani and Gupta (2017), Bardhan (2018)** and **Wang *et al.* (2018)** etc. which reported the positive and significant relation between the attitude and intention of the respondents concerned with respective behaviours.

4.3.3.2. Path 2: Subjective Norms to Behavioural Intention (SN⟶BI)

Secondly, the data revealed that the subjective norms associated with stubble burning (SN) were found to be in a positive and significant relation with the behavioural intention regarding stubble burning (BI). The estimated value of standardized path coefficient for this path was found to be 0.28 and the p-value obtained was 0.047, which is less than 0.05. Thus, this relationship was found to be significant at 5 per cent level of significance. It meant that for a unit increase in the subjective norms associated with stubble burning, there was an increase of 0.28 units in the behavioural intention of the farmers regarding stubble burning.

From the data, it could also be observed that the subjective norms were the least influential on the behavioural intention of the farmers regarding stubble burning, among the three constructs; attitude, subjective norms and perceived behavioural control, as the estimated path coefficient was the least. The results might be explained by the fact that the social norms including pressure from fellow farmers and farmer groups in the study were in favour of stubble burning. However, it was revealed during the course of the study that final decision regarding stubble burning was taken by the farmers themselves after considering various risks associated with it. This implies that the existing subjective norms were in favour of the stubble burning but that affected the intention to burn the stubble to a lesser extent than the other factors like attitude and perceived control.

Hence, it can also be noted that the null hypothesis, H_{02} was rejected, which stated that there was no significant relationship between the subjective norms and behavioural intentions regarding stubble burning. Similar results were reported in the studies like **Higuchi *et al.* (2017), Issa and Hamm (2017), Valizadeh *et al.* (2018)** and **Wang *et al.* (2018)** which stated that subjective norms had a positive and significant influence on the corresponding behaviours studied.

4.3.3.3. Path 3: Perceived Behavioural Control to Behavioural Intention (PBC⟶ BI)

The third finding drawn from the structural model estimation revealed that the perceived behavioural control associated with stubble burning (PBC) was in a positive and significant relation to the behavioural intention regarding stubble burning (BI). The standardized path estimate value of this path was 0.47. The data also revealed that p-value obtained for this path was 0.031, which was less than 0.05. Therefore, the relationship was found to be significant at 5 per cent level of significance. It implied that for a unit increase in perceived behavioural control associated with stubble burning, there was an increase of 0.47 units in the behavioural intention, which indicates a good degree of influence.

The data also revealed that it was the second most influential factor to affect the behavioural intention of the farmers, among the three TPB constructs, attitude, subjective norms and perceived behavioural control. This might be explained by the fact that the ease of burning the stubble, the affordability of the monetary fines on getting caught for stubble burning and the cumbersome procedure to apply for subsidies in order to avail machines influenced the intention of the farmers to burn the stubble to a better extent than their peers and farmer groups.

The findings also concluded that the null hypothesis which stated that there was no significant relationship between perceived behavioural control and behavioural intention regarding stubble burning, H_{03} was rejected. **Zhang *et al.* (2015), Coffie (2015), Borges *et al.* (2016), Zeweld *et al.* (2016)** and **Wang *et al.* (2018)** stated the similar findings that perceived behavioural control positively influenced the behaviours in question in the respective studies.

4.3.4. Extent of influence of Behavioural Intention on Stubble Burning Behaviour of farmers

The structural model estimation also provided the results regarding the influence of behavioural intention regarding stubble on the stubble burning behaviour of the farmers.

4.3.4.1. Path 4: Behavioural Intention to Stubble Burning Behaviour (BI ⟶ SBB)

The data in Table 43 shows that the path between behavioural intention regarding stubble burning (BI) and the stubble burning behaviour of the farmers (SBB) was estimated with the standardized value of 0.53. The obtained p-value was 0.018, which was less than 0.05. Thus, the relationship between the two variables was found to be significantly positive at 5 per cent level of significance. It meant that for each unit increase in behavioural intention regarding stubble burning, there was an increase of 0.53 units of stubble burning behaviour.

Table 43: Findings of Structural Model Estimation (II)

Path Description	Standardized Path Estimate	p-value	Hypothesis	Result
BI ⟶ SBB	0.53**	0.018	H_{04}	Rejected
BI= Behavioural Intention regarding Stubble Burning, **SBB**= Stubble Burning Behaviour, **p< 0.05				

The findings also revealed that the null hypothesis which stated that there was no significant relationship between the behavioural intention and stubble burning behaviour of the farmers, H_{04} was rejected. The present findings were found to be in line with those of **Bond *et al.* (2010), Higuchi *et al.* (2017), Issa and Hamm (2017), Maichum *et al.* (2017)** and **Sukhmani and Gupta (2017)** which stated that behavioural intentions affected the behaviours under study.

The path diagram for the structural model obtained in the Figure 32:

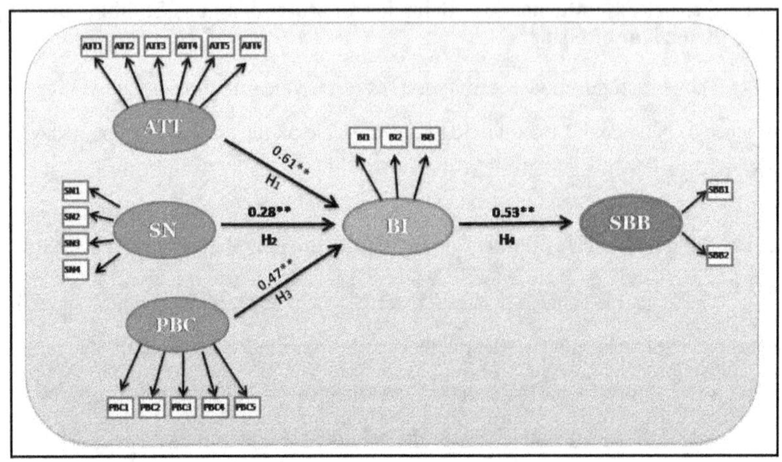

Figure 32: The path diagram for various paths in the structural model

This could be the explaining factor to the fact that most of the farmers who intent to burn the stubble, actually burnt it in the past. In the absence of a viable, economic and convenient stubble management measure, the farmers were prepared to burn the stubble in the foreseeable future as well. On the other hand, some farmers who could afford other stubble management measures showed that they might turn to them in near future as they were concerned about the harmful effects it had on soil, human health and environment as a whole. Some of the farmers were found to burn a portion of stubble instead of burning the complete left-over stubble. They only burnt the stubble on the field wherever it was necessary and managed rest of the stubble by other measures such as using Happy Seeder etc. However, during the course of investigation, it was also revealed that farmers could stop burning the stubble if provided with better alternatives. This might prove to be an important finding in the sense that if the intentions could be changed (by changing attitudes, subjective norms and facilitating better behavioural control over other stubble management measures), then there are good chances of facilitating a change in behaviour as well.

Hence, this section described the various steps involved in Structural Equation Modeling leading finally to our objectives of the study to find out the influence of various TPB constructs (attitude, subjective norms, perceived behavioural control) on behavioural intention and that of intention on stubble burning behaviour of the farmers. The section started from testing the validity and reliability of the measurement model using Confirmatory Factor Analysis by Chronbach's alpha, Average Variance Extracted and factor loadings for each TPB construct (attitude, subjective norms, perceived behavioural control, behavioural intention and stubble burning behaviour). The model was found to be adequately reliable and valid according to recommended values. Then it was tested for fitness using seven standardized fitness indices involving χ^2/df, GFI, CFI, AGFI, IFI, NFI and RMSEA. They all possessed above recommended values, thus the measurement model showed good fit. After that the structural model was tested for fitness using similar fit indices and all of them also showed good fit. Lastly, the standardized path coefficients were calculated using AMOS software for various paths (**ATT → BI, SN → BI, PBC → BI and BI → SBB**). The results showed that the standardized path coefficients for the paths were 0.61, 0.28, 0.47 and 0.53 respectively. All the hypothesized relationships between the TPB constructs were found to be significant and positive. Thus, all the null-hypotheses which stated that there was no relationship between the TPB variables were rejected. It was also inferred that attitude towards stubble burning was found to be the most influential factor governing behavioural intention of the farmers regarding stubble burning, followed by perceived behavioural control related to stubble burning. The subjective norms associated with stubble burning were found be least influential on the behavioural intention of the farmers regarding stubble burning, although they shared a significant and positive relationship. It was also concluded that the behavioural intention of the farmers regarding stubble burning also bear a significant and positive relationship.

4.4. RELATIONSHIP BETWEEN PROFILE CHARACTERISTICS AND STUBBLE BURNING BEHAVIOUR OF THE FARMERS

In order to find out the relationship between the profile characteristics and the stubble burning behaviour of the farmers, the correlation test and chi-square test were applied using Statistical Package for Social Sciences. The variables pertaining to profile characteristics were: age, gender, education, size of land holding, annual income, livestock possession, type of farming, information seeking behaviour, innovativeness, risk orientation, scientific orientation, ecological consciousness, economic motivation, awareness about stubble management measures and cropping intensity.

The summary of the results of the correlation co-efficient between the profile characteristics of the farmers and their stubble burning behaviour with their significance value is shown in Table 44.

Table 44: Relationship between profile characteristics and Stubble Burning Behaviour of farmers

S. No.	Variables	Calculated correlation coefficient values "r"	Calculated significance value "p"
1.	Age	0.084^{NS}	0.094
2.	Size of land holding	0.006^{NS}	0.926
3.	Annual Income	-0.008^{NS}	0.895
4.	Livestock possession	0.046^{NS}	0.474
5.	Information seeking behaviour	-0.099^{NS}	0.122
6.	Innovativeness	0.015^{NS}	0.817
7.	Risk Orientation	0.011^{NS}	0.864
8.	Scientific Orientation	0.045^{NS}	0.848
9.	Ecological consciousness	0.001^{NS}	0.986
10.	Economic Motivation	0.495^{*}	0.023
11.	Awareness regarding alternative stubble management measures	0.012^{NS}	0.260
12.	Cropping intensity	0.041^{NS}	0.518

NS: Non-Significant
* Correlation is significant at the 0.05 level.

The table reveals the relationship between the variables related to profile characteristics and stubble burning behaviour of the farmers. It was found that the relationship between profile variables like age, size of land holding, livestock possession, innovativeness, risk orientation, scientific orientation, ecological consciousness, awareness regarding alternative stubble management measures, cropping intensity, annual income and information seeking behaviour; and stubble burning behaviour of the farmers was non-significant. The strength of the correlation between these variables and stubble burning behaviour of the farmers was very weak, denoted by a very less 'r-value' and 'p-value' was more than 0.05 in case of each variable. Thus, hypotheses which stated the relationship between these variables and stubble burning behaviour, H_{05}, H_{06}, H_{07}, H_{08}, H_{09}, H_{010}, H_{011}, H_{012}, H_{013}, H_{015}, H_{016} were accepted.

It was found that economic motivation was the only variable to have a positive and significant relationship with the stubble burning behaviour of the farmers. The 'r-value' in case of economic motivation was 0.495 which reflects its moderate strength of relationship with the stubble burning behaviour of the farmers. The observed 'p-value' for economic motivation was 0.023, which was less than 0.05, indicating that the relationship was significant. Hence, null hypothesis, which stated that "There is no significant relationship between economic motivation of the farmers and their stubble burning behaviour", H_{014} was rejected. As most of the farmers were having high economic motivation, this might be one of the key factors in choosing a stubble management measure. Since, stubble burning is the most economical measure available, most of the farmers preferred it instead of going for other alternatives. This also helps in deducing the reason behind farmers reverting back to burning after trying the alternative measures as they were costly and didn't yield fruitful results.

On the other hand, the relationship between variables with grouped data like gender, education and type of farming; and the stubble burning behaviour were analyzed through chi-square test. The results of chi-square statistics for these variables are given Table 45, 46 and 47 respectively. The data shows that in case of all the three variables; gender, education and type of farming, chi-square calculated values were less than chi-square tabulated values (χ^2 cal $<$ χ^2 tab) at respective degrees of freedom. Therefore, they were independent of stubble burning behaviour of the farmers. Hence, hypotheses H_{017}, H_{018} and H_{019} were accepted.

Table 45: Relationship between Gender and Stubble burning Behaviour

			Gender		Total	χ^2 value
			Female	Male		
Stubble Burning Behaviour		High	13 (11.8)	119 (120.2)	132 (132)	χ^2 (cal)= 3.903
		Low	1 (4.4)	48 (44.6)	49 (49)	
		Medium	8 (5.8)	57 (59.2)	65 (65)	χ^2 (tab) = 5.991
Total			22 (22)	224 (224)	246 (246)	

Degrees of freedom = 2
*Expected count is given in brackets

Table 46: Relationship between Education and Stubble Burning Behaviour

			Education							Total	χ2 value
			Illiterate	Functionally literate	Primary education	High School	Intermediate	Graduate	Post Graduate		
Stubble Burning Behaviour	High		8	19	15	24	27	30	9	132	
			(9.1)	(16.1)	(17.2)	(25.2)	(29)	(28.4)	(7)	(132)	χ2 (cal) =8.720
	Low		5	5	7	13	8	9	2	49	
			(3.4)	(6)	(6.4)	(9.4)	(10.8)	(10.6)	(2.6)	(49)	
	Medium		4	6	10	10	19	14	2	65	
			(4.5)	(7.9)	(8.5)	(12.4)	(14.3)	(14)	(3.4)	(65)	χ2 (tab) =21.026
Total			17	30	32	47	54	53	13	89	
			(17)	(30)	(32)	(47)	(54)	(53)	(13)	(89)	

Degrees of freedom = 21

*Expected count is given in brackets

Table 47: Relationship between Type of farming and Stubble Burning Behaviour

		Type of farming			Total	χ^2 value
		Diversified	Mixed	Specialized		
Stubble Burning Behaviour	High	37 (33.8)	34 (37.6)	61 (60.6)	132 (132)	χ^2 (cal)= 4.661
	Low	7 (12.5)	17 (13.9)	25 (22.5)	49 (49)	χ^2 (tab)= 9.448
	Medium	19 (16.6)	19 (18.5)	27 (29.9)	65 (65)	
Total		63 (63)	70 (70)	113 (113)	246 (246)	

Degrees of freedom = 4
*Expected count is given in brackets

4.5.   SOCIAL MARKETING PLAN

The present research aimed at studying the stubble burning behaviour and its determinants based on Theory of Planned Behaviour (TPB). The previous sections have covered the profile characteristics of the farmers, their stubble burning behaviour and also the extent of influence of TPB constructs (attitude, subjective norms and perceived control) on the behavioural intention and that of intention on stubble burning behaviour of the farmers. This has led to the final objective of the study, i.e. planning an intervention by which the stubble burning behaviour of the farmers could be averted. Since, herculean efforts have been made by various governments and non-government agencies using legislative and judicial measures; still the problem of stubble burning seems to be far from over. In such a case, focus need to be shifted to the attitudinal and behavioural aspects of the problem. The present study has reported the factors that contribute to the stubble burning behaviour of the farmers. The only logical end to this study is by crafting a plan that could help making change in the behaviour of the farmers based on those factors. In same lines, i.e. in the context of changing behaviours in diverse fields, social marketing has earned a reputation over the years, across the globe.

Social marketing involves the application of the marketing discipline to social issues and causes. It provides a framework for developing innovative solutions to social problems that have long been a concern to the society. It has emerged from business marketing practice as a social change tool uniquely suited to achieve social profits by designing integrated programs that meet individual needs for moving out of poverty, enabling health, improving social conditions and having a safe and clean environment. There is enough evidence in the review which shows that social marketing can form an effective framework for behaviour change interventions and can provide a useful 'toolkit' for organisations or agencies that are trying to change various kinds of behaviours related to health, environment, etc. (Ross *et al*, 2006).

Consequently, a social marketing plan has been proposed using BEHAVE framework given by Academy of Educational Development (2008) and utilizing the relevant findings drawn from the stubble burning behaviour of the farmers and its determinants found in the study area. The BEHAVE framework has proved significant in inducing behaviour change across varied social issues from public health to environmental protection, to traffic safety to education reforms (AED, 2008).

The social marketing plan has also taken into consideration the socio-personal, economic, psychological and communication characteristics of the farmers for the purpose of segmentation of the target audience. Since, the present objective is part of the requirement for doctoral dissertation, it holds resource and authority-related constraints, and therefore, is limited to drawing needed strategic recommendations. These recommendations have been made in view of the expressed opinions of the farmer respondents, observations of the investigator and the inferences drawn from the study.

The findings of the study revealed that all the determinants of behavioural intention regarding stubble burning (attitude, subjective norms and perceived behavioural control) influenced it to some extent. However, attitude of the framers towards stubble burning was the most influential factor in affecting the behavioural intention of the farmers. So, the proposed social marketing plan consists of two parts; the first part addresses attitude change and the second part concern the overall stubble burning behaviour, as changing the attitudes is highly likely to facilitate a change in behaviour. So, the two components of the social marketing plan constructed using BEHAVE framework are as follows:

4.4.1. Plan focussing on the attitude change

The findings in the previous sections revealed that majority of the respondents (51.63%) have positive attitude towards stubble burning, followed by farmers with neutral attitude (35.37%) and negative attitude (13.00%) towards stubble burning. The findings in the previous section also pointed that out of the six items of the attitude towards stubble burning, the factor loading of the item ATT1 (Burning is the most economic stubble management measure) was found to be 0.932, which was the highest. Since, a factor loading value close to 1 signifies that the factor strongly influences the variable. Therefore, the economic aspect of the alternative measures needs to be addressed. On the other hand, it was also revealed that the factor loading of the item ATT3 (Stubble burning causes environmental pollution) was found to be 0.851. This indicates that farmers were also aware of the potential threats caused by stubble burning on the environment. Also, the factor loading of the item ATT4 (Stubble burning is bad for human health) was recorded to be 0.781. Therefore, it can also be deduced that the farmers acknowledged the ill-effects of stubble burning on human health. However, the economics weighed over the health and environmental effects of stubble burning.

In essence, it can be concluded that in order to change the positive attitude of the farmers towards stubble burning, following aspects should be considered:

- The alternative stubble management measure should be low-cost.
- It should be environment friendly.
- It should safeguard human health.

Further, it is well established that the attitude of all the farmers cannot be changed. So, the efforts should focus on bringing the farmers with positive attitude towards stubble burning into neutral or negative attitude categories. For this purpose, following plan has been suggested:

4.4.1.1. Name the bottom line: The bottom line of this component could be "changing the attitude of the farmers regarding stubble burning by making them realise the opportunity cost of low-cost stubble management alternative."

4.4.1.2. Select the target audience: The target audience for changing the attitudes could be those farmers who burn the stubble and who could not afford the expensive mechanical measures. For this, the farmers belonging to 'very high frequency' category of stubble burning and those with semi-medium land holding category could be selected. The reason behind this was that the farmers belonging to these categories were in significant numbers and most of them showed positive attitude towards stubble burning.

4.4.1.3. Develop a strategy: The strategy for changing the attitudes should focus on the opportunity cost of the low-cost alternatives like Pusa Decomposer. The Indian Agricultural Research Institute has recently launched a product named Pusa Decomposer, which is basically a multi-microbial formulation that converts the stubble biomass into organic matter in a short period of time. More details about this product is provided in the subsequent sections. The opportunity cost refers to the potential benefits an individual misses out on, when choosing one alternative over another. In the context of present study, the respondents even after being aware of health and environmental threats posed by stubble burning, preferred it. This indicates that stubble burning by the farmers was chosen at the opportunity cost of health and environment. However, by using Pusa Decomposer, which comes at a price of just Rs. 20 for an acre, the farmers can reap the benefits of better health and safer environment. The cost of Rs. 20 per hectare is affordable for all the farmers. Also, the benefits which it brings in the form of improved soil health, better yields etc. certainly overweigh the losses incurred by the farmers in a long run.

The next consideration should be how this message could be transferred among the target audience in an effective manner. Persuasion has been defined as "the process by which a message induces change in beliefs, attitudes, or behaviours" (Myers, 2011). Studies have shown that the credibility of a perceived message has been found to be crucial factor in persuading the target audience. So, the use of credible sources like fellow farmers or progressive farmers could be made in the context of present study. Secondly, the nature of the message plays an important role in persuasion. Sometimes presenting both sides of a story is useful to help change attitudes. When people are not motivated to process the message, the number of arguments presented in a persuasive message has been found to influence attitude change, such that a greater number of arguments will produce greater attitude change. So, there is a need to focus on both the aspects, that is, pros of using low-cost alternative stubble management measures and the cons of stubble burning to the farmers. Thirdly, a message can appeal to an individual's cognitive evaluation to help change an attitude. This can be done in one of the two ways. Either the target individual is presented with the data and motivated to evaluate the data and arrive at an attitude changing conclusion. While in the second way, the target individual is encouraged to look not at the content but at the source. This is commonly seen in modern advertisements that feature celebrities. So, a mix of all these persuasive techniques may be used in the communication strategy. In short, the farmers must be made aware of the benefits they are missing-out on by burning the stubble and how Pusa Decomposer can help them reap those benefits.

4.4.1.4. Select a media mix: The next step should be selecting a media mix that can be used to convey the messages highlighting the benefits of Pusa Decomposer to the farmers. The findings of the study revealed that television, mobile phones and newspapers were the prevalent mass media sources used by the farmers in the study area for gathering farm related information. Hence, these channels could be utilized in a planned and systematic manner. The advertisements promoting the use of Pusa Decomposer could be run on television channels at evening or night hours, when farmers usually get back home after work. Farmers were also found to be regularly getting text SMS alerts from various agencies regarding agricultural information. These could also be well utilized to draw the attention of the farmers towards the product. Apart from these, many farmers were active on social media platforms like WhatsApp, Facebook and YouTube. The farmers used these platforms to share agricultural information in their groups. The video messages featuring method and result demonstrations of the application of Pusa Decomposer could be floated on these platforms. The advertisements and success stories regarding use of Pusa Decomposer could also be published in newspapers so that the message reaches a large audience.

Thus, based on the findings of the study, messages could be designed in various formats like text messages, videos, digital images/posters (mobile phones), advertisements and flash headlines (TV) and articles like success stories, features etc. (newspapers). Further, for the easy interpretation by the respondents, the messages in all the formats should be designed in regional language.

4.4.1.5. Pre-testing of the messages: The messages designed for the promotion of the product needs to be pre-tested among a small representative target audience via pilot testing. Careful attention should be provided to test the messages for appropriate interpretation by the target audience and prevent ambiguity. The necessary modifications should be made. The media mix selected should also be checked for proper working in this stage.

4.4.1.6. Implementation: Planning the things out is only half the job done. Every aspect planned in the previous stages needs to be brought to action. The lessons learnt during pre-testing should be incorporated and the continuous flow of messages should be ensured during this stage.

4.4.1.7. Evaluation: The evaluation criteria for measuring the attitude change should be set in advance. Appropriate tools and techniques should be constructed before the implementation of the plan. The change in the attitude of the respondents could be measured using an attitude-scale by evaluating them on pre-test and post-test basis. It should be ensured that the evaluation process takes place when it is scheduled.

4.4.1.8. Refining: After the evaluation, the areas which need any kind of rectification must be addressed appropriately. The schedule for refining should also be set in advance and adhered properly.

The Figure 33 shows the social marketing plan focussing on the attitude change:

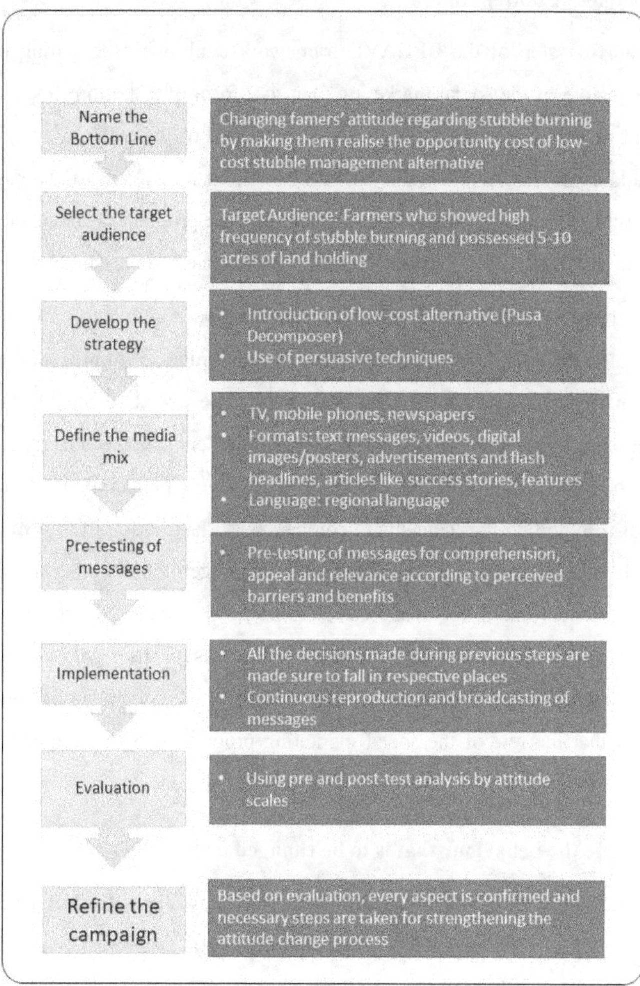

Figure 33: Proposed social marketing plan (attitude component) using BEHAVE framework

4.4.2. Plan focussing on behaviour change

Once the change in attitude is noticed, then changing the behaviour is supposed to become easy. The second component of the social marketing plan focuses on the behaviour change. The BEHAVE framework will be followed for this component as well:

4.4.2.1. Name the bottom line

The first step of the BEHAVE framework deals with the setting up of the overall objective of the social marketing plan in a form of a 'bottom line'. This first step ought to hold the background, focus and purpose of the entire social marketing plan. It addresses the social benefit expected to occur as a result of the program. It will also serve as a judging point for those funding the program or at least those who will be judging its success. It should be something simple and measurable.

In the context of present study, the focus and purpose of the social marketing plan is to alter the stubble burning behaviour of the farmers, or to prevent the farmers from burning the paddy stubble after harvesting. The extent of stubble burning is often measured by the number of stubble burning incidents observed or reported in an area. So, the bottom line of the proposed social marketing plan could be, "Reduce the number of stubble burning incidents in comparison to last season in Malwa region of Punjab." It is to be noted that the number or percentage of stubble burning incidents should be set according to the available resources for developing the social marketing program. However, it should be large enough to measure the significant behaviour change among the target audience, i.e. the farmers. By clearly stating this goal, it can be ensured that the rest of the social marketing program will be designed with this purpose in mind.

4.4.2.2. Name the behaviour that is to be changed

A behaviour refers to a specific action taken by a specific audience under a specific set of circumstances. In social marketing, a target behaviour in social marketing consists of two parts: the 'actor' and the 'action'. The two are inter-connected. One cannot be addressed without the other. In this step, the exact behaviour which is needed to be changed is defined and the target audience for which this social marketing plan is being prepared, is to be pin pointed. The behaviour under concern in the present study is burning the paddy stubble after harvesting by the farmers. In the previous sections, various profile characteristics of the sampled farmer respondents have been described. The summary of both, the behaviour (action) and the target audience (actors), is as follows.

The stubble burning behaviour was studied using two components, i.e. frequency of engagement in stubble burning behaviour and the proportion of the total produced paddy stubble managed through burning. The 'frequency' referred to the number of occasions on which the farmer burnt the stubble on field after harvesting in past five years. On the other hand, the 'proportion' referred to the part of the total produced stubble which was managed through burning, in case other stubble management methods were also used by the farmer. It was found that majority of farmer respondents were engaged in stubble burning every year in the last five years, thus classified under 'very high frequency' (58.13%). Corresponding to the 'proportion' of the stubble burnt, it was found that the most number of respondents (49.19%) burnt more than 80 per cent of the total stubble produced on their field and were classified under 'very large' proportion category. The data regarding overall stubble burning behaviour revealed that the majority of farmer respondents showed 'high' performance (53.66%) followed by 'medium' (26.42%) and 'low' (19.92%) performers respectively.

The findings regarding socio-personal characteristics of the respondents revealed that most of the respondents were middle-aged (65.46%) between 26-50 years, male (91.06%), had formal education up to intermediate level (21.95%) and possessed semi-medium land holding (27.24%), i.e. between 5-10 acres. The results related to economic characteristics revealed that majority of the farmers (51.46%) belonged to low annual income category i.e. in between Rs. 50,000 to Rs. 4,50,000, possessed lager milch animal like cows and buffaloes (98.78%) and were practising specialized farming (45.93%).

Based on the above findings, the proposed social marketing plan will cater to those farmers who showed very high frequency of stubble burning and possessed semi-medium land holding, i.e. between 5-10 acres. The reason behind the selection is based on the numbers. The 'very high performance' of the stubble burning behaviour shown by the farmers in terms of frequency will be reflected in the number of stubble burning incidents, on which our whole social marketing plan is based. On the other hand, the number of farmers falling in semi-medium land holding category was found to be the highest. It was also noted that the farmers who had bigger land holdings tend to at least try other stubble management measures like using machines e.g. Happy

Seeders, Super Seeders etc. and were able to bear extra expenses that occurred in implementing those alternative measures. So, the idea is to target those who cannot afford much of the extra expenses and thus, prefer to burn the stubble. It can also be argued that the farmers falling under low-income categories could have been selected as the target audience. But it was noted that most of the farmers bore both characteristics (semi-medium land holding and low income) and it would be easy to locate the farmers with the land holdings. Hence, the selection was made as such.

4.4.2.3. Develop a strategy

Once the behaviour is clearly stated and the target audience is identified, the next step involves formulating what might be done do to change the behaviour under concern. In this step, after conducting formative research, analysing the results, specific determinants of the concerned behaviour are identified. Under BEHAVE framework, these determinants are known as 'barriers' and 'benefits'. Barriers refers to the reasons the target audience cannot (easily) or does not want to adopt the behaviour. On the other hand, 'benefits' are the reasons the target audience might be interested in adopting the behaviour or what might motivate them to do so. Apart from these, there is another essential component to be addressed, i.e. the 'competitors' or the 'competing behaviours' that target audience prefers over "desirable" behaviours. An analysis is done for both, the 'desired behaviour' and the 'competing behaviour' regarding barriers and benefits. This helps in deciding how to increase the benefits or decrease the barriers of the desired behaviour and reduce the benefits or increase the barriers of competing behaviour. This step ought to end in a summary, i.e. a strategy which should be expressed in three to four easy-to-remember bullet points. Or better yet, the strategy can also be boiled down to a single declarative sentence, if possible.

In the context of present study, the 'competing behaviour' refers to 'burning the stubble' while the 'desired behaviour' refers to 'reducing the stubble burning incidents'. The findings and discussions in above sections of the chapter have revealed that there were very less barriers to the competing behaviour. During the investigation, it was revealed that there were provisions for monetary fines and even imprisonment on getting caught for burning the stubble but hardly anyone was ever imprisoned. The monetary penalties were also not hefty, and farmers find it easy to get away with them due to their affordability. The lack of constant monitoring also adds to the convenience of the farmers to burn the stubble. However, these restrictions keep on getting imposed and lifted from time to time. Another barrier was noted to be the ill effects of stubble burning on the soil health. Some of the farmers were aware of the dire consequences but still they preferred to burn due to various reasons. Some of the farmers also expressed that they, along with their family members faced breathing problems, eye-irritation, allergies, etc. during the burning season, which often caused problems to old, sick and children. However, this was not the case for all the farmers. So, overall, there exist these barriers for burning the stubble, from farmers' point of view.

On the other hand, the number of benefits seemed to outnumber the barriers in case of competing behaviour. The farmers reported several benefits like burning the stubble was the cheapest measure available to them. It was a lot convenient to simply burn the stubble than to apply for subsidies for machines or put extra money on implements to manage the stubble in-situ. Other measures like zero tillage did not satisfy most of the farmers because of weed infestation and reduced germination in next crop. Burning the stubble also saved a lot of time in the field preparation and consequently sowing of the wheat crop. All these factors were also reflected in the formation of positive attitude of farmers towards stubble burning and perceived behavioural control related to stubble burning. In the previous sections, it was also revealed that the attitude and perceived behavioural control influenced the behavioural intention regarding stubble burning in a positive and significant manner, which in turn determine the stubble burning behaviour. So, all these add to the credentials of competing behaviour.

The desired behaviour, on the other hand, faced multiple barriers that prevented the farmers from not burning the stubble. First among these were the high cost of alternatives available. The available alternatives included use of machines like Happy Seeder, Super Seeder, in-situ incorporation, bale making, etc. All these measures added weight to the already burdened pockets of the farmers. The extra fuel costs, maintenance costs of the tractors, time, energy, labour, etc. incurred in the extra tillage operations proved to be too much for most of the farmers. There exists a provision of subsidies, but farmers expressed their displeasure in that aspect also due to cumbersome procedure and distrust on the middlemen who ought to help in managing those subsidies. All these factors add up in formation of negative attitude of the farmers towards available alternatives.

On the contrary, looking at the brighter side, the desired behaviour tends to provide farmers rid of health issues that arise particularly in stubble burning season and improve their field's soil fertility. According to the latest legislative measures, the farmers are to be provided with monetary compensation in case they do not burn the stubble on their fields. Lastly, it is going to reduce the pollution levels, for which the farmers are arguably held as the solid reason for because of burning the stubble, and thus, providing them with a sense of satisfaction and relief. So, the number of benefits may look more in comparison to the barriers in case of the desired behaviour, but it needs to be brought to the notice of the target audience. Not only just bring to their notice but also be tried to inculcate into their attitudes which could ultimately facilitate their behaviour change. The Table 48 indicates the benefits and barriers of both kinds of behaviour.

Table 48: Determinants of desired and competing behaviours

	Competing Behaviour	**Desired Behaviour**
Barriers	<ul><li>Monetary fines (which keeps on imposing and lifting)</li><li>Hampering soil fertility</li><li>Health issues</li></ul>	<ul><li>High cost of alternatives</li><li>Cumbersome procedure to apply for subsidies</li><li>Negative attitude towards alternatives</li></ul>
Benefits	<ul><li>Economical</li><li>Convenient</li><li>Time saving for sowing of wheat crop</li><li>Easy field preparation for next crop</li></ul>	<ul><li>Health benefits</li><li>Improvement in soil health</li><li>Rewards/ Compensation</li><li>Reduction in pollution</li><li>Satisfaction</li><li>Relief</li></ul>

Apart from these barriers and benefits, the investigation also featured the viewpoints of farmers regarding how the stubble burning could be stopped. As mentioned earlier also, most of the farmers were ready to stop burning if a low-cost alternative is made available to them. Some of the farmers suggested that stubble burning could be stopped if the produced stubble is consumed by factories, industries etc. directly from the farmers' fields so that farmers do not have to incur extra cost of transportation to move it outside the fields. The findings showed that majority of the farmers (67.48%) were falling in the medium category of cropping intensity, i.e. they were growing 2-3 crops on their fields in a year. Out of these crops, paddy and wheat were the major crops. Some farmers opined that if government could provide better Maximum Selling Price of other crops like mustard, cotton, pea etc. which can be grown successfully in the area, they could shift to those crops as well. This will also reduce the rate of declining water table due to continuous growing of high water-requiring crops like paddy in the area. Some of the farmers who were against the stubble burning expressed that hefty monetary penalties should be awarded to those who burn the stubble and high compensation or incentives to those who do not burn the stubble. There were also some farmers who felt the need of awareness about the ill-effects of the stubble burning in their areas.

So, keeping in mind the above suggestions provided by the farmers and the 'bottom line' of our proposed social marketing plan, our strategy should focus basically on following points:

- **Introduce a low-cost alternative stubble management measure**

- **Provide a good compensation/incentive to those who do not burn the stubble**

- **Use emotions to remind the farmers of ill-effects of stubble burning at regular intervals using various methods**

- **Make the farmers aware of benefits of not burning the stubble**

4.4.2.4. Define the marketing mix

Once the benefits and barriers of the behaviour according to audience's perspective are understood, it is time to construct the marketing mix. Kotler (2000) defined marketing mix as the controllable group of variables that the organization can use to effect on the buyer's response to product or service. It basically features 4 Ps of marketing, namely Product, Price, Place and Promotion. With the time, various authors and experts have added other Ps such as policy, partnership, publics, purse strings etc. to the mix. As important the marketing mix is to the commercial marketing, it is equally essential to the social marketing as well.

The step by step description of the 4 Ps is given below:

(i) Product: The decision about selecting the product or services is always the first step. The "product" is expected to reduce the barriers and increase the benefits of the desired behaviour. It is not mandatory for the 'product' to be a physical or tangible entity. It could be intangible such as a service, an idea, or an action. There are many aspects of product to be considered before it gets launched among the target audience. A product or service ought to have features like function, appearance, packing, and guarantees of performance that help people solve problems. Other aspects that need to be considered are frequency of the product to be used, the effectiveness of the product in solving the problem, competition in the market, availability of other alternatives and the durability of the results delivered by the product.

In the context of present study, among the dearth of low-cost alternatives to stubble burning, the Microbiology Division of Indian Agricultural Research Institute, Indian Council of Agricultural Research (ICAR-IARI), New Delhi has developed a "microbial consortia of eight types of micro-organisms", known as "Pusa Decomposer" (Figure 34). It contains strains from fungi, which assist in producing enzymes essential to quicken the decomposition of biomass (stubble). The decomposer should be fermented for nearly a week. The solution, for preparing which other organic inputs like jaggery and chickpea flour are also used, is then sprayed over the fields. It reduces the decomposition time of shredded and watered paddy stubble from around three months to just 25 days. Applying Pusa Decomposer retains a portion of the stalk and the roots, which get decomposed in the fields themselves and add to the nutrition in the soil. This helps retain several tonnes of nutritious residue in the form of compost per every hectare of land.

One packet of Pusa Decomposer contains four capsules which is enough for one hectare of land. It is completely organic and chemical free, thus eliminating the chances of being poisonous. It was prepared after more than five years of research and extensive field trials in states of Punjab, Haryana, Rajasthan, Uttar Pradesh and Delhi. It has been recently introduced in late 2020 and thus, was not a part of data collection process. However, it has been promoted by the government agencies and even recommended to the farmers across the northern states. Reports also suggest that IARI has been closely working with Punjab Agricultural University to propagate it in the state of Punjab. Eight multinational companies have already been tied up for its bulk production for next year.

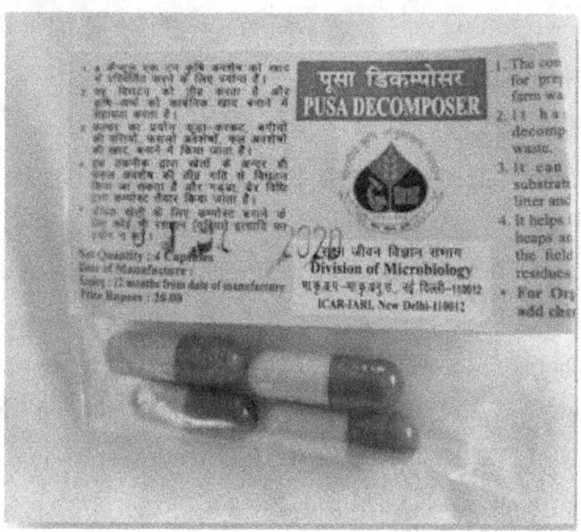

Figure 34: Pusa Decomposer, the proposed Product for the social marketing plan

Looking at the desirable qualities in a social marketing product, Pusa Decomposer is low-cost (Rs. 20 per hectare), of low frequency use (one-time use in a year), very cost-effective from other costly mechanical measures and easy to use. Hence, it is the recommended product for the proposed social marketing plan. However, the product is to be provided for free trials to the farmers and promoted among the farmer groups, which is reported to be happening. The feedback of the farmers should be regularly taken so that the farmers feel connected and the transition of the product from the shop to the field is smooth. So, many efforts are already in progress to diffuse this product in the market.

The findings of the present study revealed that majority of farmers possessed medium-high levels of innovativeness (77.24%), scientific orientation (82.67%) and risk orientation (77.67%). It was also found that maximum number of respondents showed high level of ecological consciousness (45.53%). So, it is highly expected of them to go with this alternative to stubble burning which is new, scientifically recommended, nature friendly and involves minimal risk.

(ii) Price: "Price" refers to what the consumer must do in order to obtain the social marketing product. This cost may be monetary, or it may instead require the consumer to give up intangibles, such as time or effort, or to risk embarrassment and disapproval. If the costs outweigh the benefits for an individual, the perceived value of the offering will be low, and it will be unlikely to be adopted. However, if the benefits are perceived as greater than their costs, chances of trial and adoption of the product is much greater.

In the context of present study, the price of the product, i.e. Pusa Decomposer is placed at Rs. 20 for 4 capsules, which is considered enough for working in an area of one hectare. However, to learn about its preparation and application, farmers may have to invest some of their precious time in a short-term training as well. In addition to this, the preparation of the final solution requires jaggery or chickpea flour during the fermentation process. These inputs are usually available on farm or at home, however for the sake of calculations, it would add a maximum of Rs.50 to the total cost of the product to be used in a hectare. Since our target group is of farmers having land holding between 5-10 acres, i.e. it will effectively cost them up to Rs. 140-280 for their entire land holdings for a season. Also, if the jaggery and chickpea flour are available at home, then it will cost just Rs. 40-80 for their entire land holdings for a season.

Now, looking at the returns, the Pusa Decomposer is estimated to convert a hectare land's biomass into organic matter, which will result in soil fertility and eventually the yield of the crops will be better. So, at the expense of Rs. 100, farmers can earn even in thousands. This clearly indicates that for the target audience (the farmers), the benefits will exceed the costs in a long run.

As the economic motivation of most of the farmers was found to be high (42.28%) in the present study, there are high chances that they would turn to this economical measure to manage the paddy stubble instead of burning.

(iii) Place: "Place" describes the way that the product would reach the consumer. For a tangible product, this refers to the distribution system-including the point of outlets where it is sold, or places where it is given out for free. For an intangible product, place is less clear-cut, but refers to decisions about the channels through which consumers are reached with information, training and other such interventions. Another element of place is deciding how to ensure accessibility of the offering and quality of the service delivery. By determining the characteristics and preferences of the target audience, as well as their experience and satisfaction with the existing delivery system, the most ideal 'place' or means of distribution can be pinpointed.

In the context of present study, the information seeking sources were assessed and it was found that the majority of the farmer respondents consulted their personal cosmopolite sources for information related to agriculture. Among these personal cosmopolite sources, the input retail dealers/shopkeepers and company agents were most frequently contacted by the farmers. These sources also played a crucial role in deciding about the application of inputs (types, dosage, time of application etc.) on field and they were the main sources of latest inputs available in the market. Thus, these **input retail shops and company agents** could serve as the desired 'place' for the purchase point of the proposed product. Since the primary function of the Place P is to ensure easy access to products and services, input retail dealers/shopkeepers and company agents deem to be adequately fit the proposed social marketing plan.

(iv) Promotion: The last "P" of the marketing mix is promotion. This element is often mistakenly considered as comprising the whole of social marketing program just because of its visibility. On the contrary, it is only one piece of the whole plan. Promotion consists of the integrated use of advertising, public relations, promotions, media advocacy, personal selling and entertainment channels to sell the product. Research is crucial to determine the most effective and efficient channels to reach the target audience and provide them with the product or service concerned.

In the present study, it was found that apart from the input retail dealers/shopkeepers and company agents, most of the farmers preferred television and mobile phones as their most frequently used mass media channels to access agricultural information. The regional news channels were frequently watched by the farmers in the study area. Mobile services like SMS and use of internet was also found common among the farmers. Many farmers had subscriptions of agricultural updates from Punjab Agricultural University and private organizations such as Mahindra, TATA, etc. They received regular updates on weather and farming practices in the region. During the investigation, it was also noted that many farmers were utilizing social media platforms for sharing information in the form of texts and videos. It was found that many farmers were part of groups formed by the farmers on WhatsApp and Facebook. They shared wide range of information in these groups related to variety of topics pertaining to farming. Many farmers also expressed that they used YouTube videos to gain information on the desired topics and issues, especially for any new practice that they want to learn related to farming. So, these channels could be well utilized for the promotion of the product as majority of the target audience has the access and knowledge of how to use these communication sources.

Relatives and progressive farmers were found to be the most frequently used personal localite sources for obtaining agricultural information. Since, these sources own a high degree of credibility among the farmers, they must be utilized in the promotion of the product because there are fair chances that farmers would trust the words of one of their own. The most prevalent extension methods among the farmers for accessing the agricultural information were *Kisan Melas* and meetings organized by the farmer groups. So, these platforms could also be used to highlight the advantages of the proposed product. There can also be a short demonstration session for the farmers regarding the use of Pusa Decomposer during the meetings.

In this way, these information channels could be utilized for the promotion of Pusa Decomposer among the farmers.

(v) Additional Ps: Various authors and experts have added some other Ps to the marketing mix which also play a crucial role. Some of these are publics, partnerships, policy and purse strings. "Publics" refers to both the external and internal groups involved in the program. External publics include the target audience, policymakers, and gatekeepers, while the internal publics are those who are involved in some way with either approval or implementation of the program.

"Partnership" refers to the cooperative relationship between groups who agree to share responsibility for achieving some specific goal. Most of the social marketing issues are so complex that one agency can't make a dent by itself. There is a need to team up with other organizations to be significantly effective. The organizations having similar goals may be incorporated in the planning process and identify the ways they can work together.

"Policy" refers to the governmental policies. It has been noticed that social marketing programs can do well in motivating individual behaviour change, but that is difficult to sustain unless the environment they're in supports that change for the long run. So, conducive policies are required on which the sustainable behaviours can thrive upon.

"Purse Strings" refers to the financial resources or support. Usually funds provided by sources such as foundations, governmental grants or donations Most organizations help in operating social marketing programs.

In context of present study, the partnership between several government and non-government agencies including the research, industry, extension, etc. is required. As a matter of fact, there have been tie-ups between ICAR-IARI and NGOs in states like Haryana, Punjab and Uttar Pradesh for propagating the use of Pusa Decomposer among the farmers. Several initiatives from SAUs like Punjab Agricultural University have also been taken in sensitising the masses against the stubble burning with the help of school children and voluntary organizations working in Punjab. The legislative and judiciary bodies are also bringing policies from time to time to prevent stubble burning. So, a wide array of publics is already present there and working at their own convenience and levels. They just need to be brought together on a common platform so that maximum leverage could be derived from all their efforts.

4.4.2.5. Prototyping and pre-testing

After deciding the marketing mix, the next step is prototyping and pre-testing it. Prototyping refers to the process in which the use of product is tested simulating the real environments for which it is built. In social marketing, prototyping is used to test new products and services. In short, marketers create a series of increasingly complex mock-ups and try them out with small groups of potential consumers. Pre-testing is often used to test messages for comprehension, appeal and relevance in accordance with perceived barriers and benefits.

Pertaining to the present study, the proposed product for the social marketing plan has already been tested on fields of the farmers. Reports suggest that the ICAR-IARI has been closely working with the farmers of various states like Punjab, Uttar Pradesh and Haryana regarding the promotion of Pusa Decomposer. It has also been reported that as many as 50 farmer groups in states like Punjab, Haryana and Uttar Pradesh have been approached for its demonstration among them. So, the proposed social marketing product has cleared the prototyping stage. However, the pre-testing of the messages regarding the awareness and use of the product could be performed among the target audience for necessary modifications. For pre-testing, the farmers could be gathered in focus-groups to discuss the impact of the messages and product on their perceived barriers and benefits.

4.4.2.6. Implement

The next stage in BEHAVE framework for social marketing plan is the implementation. Once, the product goes past prototyping and messages are pre-tested, it is supposed to be ready to roll out among the target audience. Simultaneously, one more thing is to be considered, i.e. preparing the evaluation criteria for the social marketing interventions. In fact, it should be prepared before the final implementation of the social marketing plan occurs.

The implementation of the plan will feature the use of lessons learned during prototyping and pre-testing of the product and messages. All the decisions made during previous steps are made sure to fall in respective places. It also features continuous reproduction of the product and the messages that are made to create awareness. It needs to be ensured that channels that are selected for promotion are timely and regularly transmitting the messages. The adequate stocks of the product are available at the points of sale of the product or not, are another aspect to be made sure of. Continuous research is needed in this step regarding awareness, attitudes and actions of the target audience to determine the effect of the interventions.

4.4.2.7. Evaluate

After the implementation, it becomes imperative to know how the proposed marketing plan is working. Research studies suggest that it is quite possible that perhaps parts of the proposed social marketing plan are effective while others are not. Hence, an evaluation criterion of the program should be designed before it is launched. It is also made sure that this evaluation relates back to the 'bottom line' or the social benefit set in the first step of the planning process. In an ideal campaign, certain parts of the evaluation are ongoing and can be measured regularly (daily, weekly, monthly, as often as possible), so the campaign can be tweaked as it moves forward.

The proposed 'bottom line' for this social marketing plan is, "Reduce the number of stubble burning incidents in comparison to last season in Malwa region of Punjab." Currently, the Consortium for Research on Agro-Ecosystem Monitoring and Modeling from Space (CREAMS) at Indian Agricultural Research Institute (IARI), New Delhi, regularly monitors the active fire events due to stubble burning at regional scales. Various other sensors like Moderate Resolution Imaging Spectroradiometer (MODIS), Along-Track Scanning Radiometer (ATSR) and Advanced Along-Track Scanning Radiometer (AATSR), etc. are used for the same purpose by different agencies (Chhabra *et al.*, 2019). So, the final evaluation, i.e. change in the number of stubble burning incidents could be traced by using these sources. Apart from these, it is equally important to assess the awareness, attitudes and perceptions of the target audience regarding the proposed product. For these, different scales, indices and knowledge tests could be constructed beforehand. The effectiveness of the dissemination channels should also be a part of evaluation process. The sale of product units from the point of sale could be an indicator of the effectiveness of dissemination channel. The evaluation schedule should be strictly adhered for both on-going and terminal components for fruitful and timely fine-tuning.

4.4.2.8. Refine the campaign

The last stage of BEHAVE framework is the refining of the campaign, based on the results of the evaluation. For this purpose, a certain time is set aside to re-evaluate whether the plan is working or not. The refining should be based on the answers to the questions like whether the picked target audience was right, whether there is any change in their behaviour, how right were the chosen barriers and benefits. It should also address the fact that whether there are any unintended consequences of the interventions, whether the messages are getting through the target audience, etc. Everything should be checked, and necessary refining should be done. Even if the results are good, nothing is perfect. There is always a little scope of improvement. The plan can always be made stronger. Before the campaign is launched, the date for this re-evaluation is set in advanced. This is to ensure that no opportunity is missed to revisit the social marketing plan and make it better.

Hence, this section featured the construction of proposed social marketing plan based on the findings of the present study using BEHAVE framework given by AED (2008). The overall bottom line for the plan was to "reduce the number of stubble burning incidents in comparison to last season in Malwa region of Punjab." The behaviour that was focussed upon, was 'reducing the stubble burning incidents' and the target audience was farmers. The plan was prepared to cater those farmers who showed very high frequency of stubble burning performance and were categorized under possession of semi-medium land holdings (5-10 acres). Then the strategy was developed keeping in mind the identified barriers and benefits of desired and competing behaviours. The strategy involved introduction of low-cost alternatives to stubble burning, provide compensation/

incentive to those who do not burn the stubble, use emotions to remind the farmers of ill-effects of stubble burning and make them aware of the benefits of not burning the stubble.

In further steps, the marketing mix involving 4 Ps were planned. The product proposed for the present social marketing plan was Pusa Decomposer developed by ICAR-IARI, New Delhi. It contains strains of fungi which decomposes the stubble into fine organic matter, good for soil health. It is priced at Rs. 20 for 4 capsules which is sufficient for a hectare of land, thus making it affordable. Based on the findings, input retail shops and company agents were identified as ideal places for sale of the product. For promotion, television, mobile phones, social media were proposed to be utilized as they involve high engagement of the farmers. Also, progressive farmers, farmers' relatives, Kisan Melas and meetings of farmer groups could also be utilized in promotion as farmers were found to depend on these sources for information.

This is to be followed by prototyping and pre-testing of the product and the messages. Reports suggest that the product has already crossed the prototyping stage. However, the messages could be pre-tested along the lines of perceived benefits and barriers. Then comes the implementation stage in which all the plans are put to action. The evaluation plan is prepared keeping in mind the 'bottom line' set in the first step. At present, the number of stubble burning incidents is traced by Consortium for Research on Agro-Ecosystem Monitoring and Modelling from Space (CREAMS) at Indian Agricultural Research Institute (IARI), New Delhi. So, that could be used for overall evaluation of the social marketing plan. it is equally important to assess the awareness, attitudes and perceptions of the target audience regarding the proposed product. For these, different scales, indices and knowledge tests could be constructed beforehand. At last, the refining needs to be done on the basis of the results of the evaluation for making the efforts to create behaviour change stronger.

The social marketing plan proposed for the study has been presented in Figure 35:

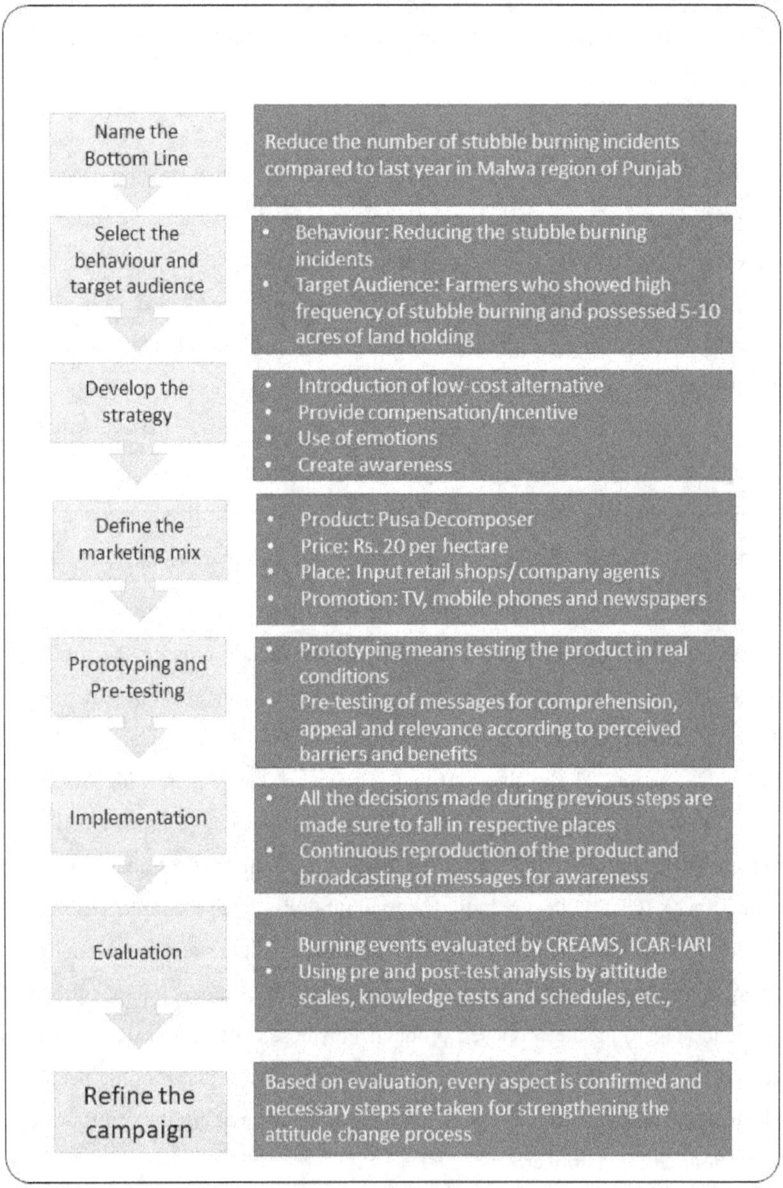

Figure 35: Proposed social marketing plan (behaviour component) using BEHAVE framework

Summary and Conclusion

One of the most prominent cropping systems in India is the rice-wheat cropping system. The Indo-Gangetic plains in Northern India accounts for around twelve million hectares of rice and wheat crop rotation. The harvesting of these crops with combine harvesters is very popular with the farmers in this region. The mechanized harvesting leaves a large quantity of stubble that takes a lot of time to decompose in the soil and is usually burnt for speedy plantation of the next crop. India generates around 600 Mt of stubble which includes about 90-140 Mt of surplus stubble that is likely to be burnt on the field (Jain *et al.,* 2014*).* Multipurpose use of stubble includes animal feeding, soil mulching, bio-manure, thatching for rural homes and fuel for domestic and industrial use. Despite knowing of alternative methods of stubble management, farmers burn a significant portion of the crop stubble on-farm so that the succeeding crop can be sown on a cleared field. Mechanized farming coupled with lack of availability of farm labour and high cost associated with the process further aggravates the problem of stubble burning. According to a study conducted by International Food Policy Research Institute, air pollution due to stubble burning in northern India causes an estimated economic loss of around USD 30 billion annually, and is a leading cause of acute respiratory infections, especially among children (Chakrabarti *et al.*, 2019). The particulate matter emitted from crop burning across India in a year is more than 17 times the total annual particulate pollution in Delhi from all sources such as vehicles, garbage burning and industries. In Punjab, total cereal stubble generated is 45.58 million tonnes/year and residue burned is 21.32 million tonnes/year (Yadav *et al.*, 2017). In 2019, the System of Air Quality and Weather Forecasting and Research (SAFAR) by the Ministry of Earth Sciences (MoES) reported that the share of stubble burning in air pollution in Delhi-NCR rose to about 46 per cent and thus, public health emergency was declared in the region.

The governments at different levels have attempted to restrict the stubble burning through numerous measures and campaigns designed to promote sustainable management methods. The National Green Tribunal banned stubble burning in 2015 but it had little or no effect in northern states. The charging of monetary penalties from the farmers who burn the stubble and FIRs against them have also not yielded fruitful results. In spite of the provision of the subsidies for mechanized measures, they are not reaching the farmers due to bureaucracy and other factors. However, it would be interesting to see the latest direction of the apex court to the Punjab government to provide incentives at the rate of Rs. 100 per quintal to the farmers who do not burn the stubble in the upcoming seasons. As of today, despite enormous efforts and measures, the desirable results to bring change in farmers' behaviour to stop stubble burning is still awaited. It seems that the problem of stubble burning is difficult to be solved by technical or legal interventions alone.

Research studies have shown that simply trying to tell people to change, or giving them information and expecting them to act on it, may not work if the determinants of certain behaviours are not considered. There is a need to sensitize the farming community by using interventions which can enable farmers to rethink and change their behaviour by forming a new perspective in this regard. In order to understand and change behaviours, social scientists rely heavily on appropriate behavioural theories to identify determinants of a particular behaviour. The present study utilized the Theory of Planned Behaviour in Indian context to the problem of stubble burning in order to understand which behavioural determinants are leading farmers to burn the stubble. Also, knowing why farmers behave the way they do is vital to the success of social marketing interventions which consequently result in behaviour change.

Therefore, keeping this in view, a study entitled **"A study on Stubble Burning Behaviour of farmers in Punjab"** was undertaken. The specific objectives of the research were as follows:

1. To study the socio-personal, economic, communication, psychological and situational characteristics of the farmers

2. To measure the stubble burning behaviour of the farmers

3. To determine the extent to which attitude, subjective norms and perceived behavioural control influence the behavioural intention of the farmers regarding stubble burning

4. To find out the extent to which behavioural intention of the farmers regarding stubble burning influences their stubble burning behaviour

5. To determine the relationship between the profile characteristics with stubble burning behaviour of the farmers

6. To propose a social marketing plan to tackle the stubble burning behaviour of the farmers

The present investigation was carried out in Malwa region of Punjab based on the maximum number of stubble burning incidents reported in the region. The descriptive research design was adopted to fulfil the objectives of the study. The Multi-stage sampling procedure was followed. Out of total eleven districts in Malwa region, districts Bathinda, Sangrur and Ludhiana were selected using simple random sampling. One block from each sampled district, viz. Bathinda block from Bathinda district, Sunam block from Sangrur district and Khanna block from Ludhiana district; were selected randomly. Two villages namely *Khialiwala* and *Amargarh* from Bathinda block; *Goh* and *Lalheri* from Khanna block; and *Lakhmirwala* and *Kharial* villages from Sunam block were selected randomly for the study. The Cochran's formula was used to estimate the sample size of the respondents for the study, which came out to be 246. The selection of respondents was done on the basis of proportional allocation method and total 246 farmer respondents were selected using simple random sampling without replacement from the selected villages. The quantitative data was collected through semi-structured interview schedule, pre-tested and modified accordingly on the basis of pilot study. Besides, secondary data, statistical facts and information relevant to the present investigation in the form of annual reports, progress report, etc. were collected from the official websites of various government and non-governmental organizations. The data collected were coded, tabulated, analysed and interpreted with the help of appropriate procedures and statistical techniques such as AMOS and SPSS software.

5.1. Major findings of the study

The salient findings of the investigation have been summarized as follows.

5.1.1. The profile characteristics of the farmers

➢ Majority of the respondents (65.46%) residing in the sampled villages belonged to the middle age category (26-50 years). There were only 19.10 per cent farmer respondents who belonged to the old age category (>50 years) followed by 15.44 per cent of the farmers who were in the young age category (<26 years).

> Majority of the respondents were found to be men (91.06%) and only 8.94 per cent of the total farmer respondents were women.

> Maximum percentages of the respondents (21.95%) had education up to Intermediate level followed by 21.54 per cent respondents had graduate level of education. Among all the respondents, 19.10 per cent had formal education up to high school level. Around 13 per cent of the farmer respondents had received formal primary education while 12.20 per cent farmers were found to be functionally literate. Only 5.29 per cent of the farmer respondents were Post Graduates while 6.91 per cent farmers were illiterate.

> Maximum number of respondents (27.24%) possessed semi-medium size, i.e. 5-10 acres of landholding. Farmers with small-sized landholding (2.5-5 acres) accounted for 21.95 per cent, followed by the farmers possessing medium sized landholding (19.51%), i.e. 10-25 acres. The marginal size of land holding (less than 2.5 acres) was held by 18.70 per cent of the farmer respondents, while 12.60 per cent of the farmers owned large size landholdings, i.e. more than 25 acres.

> Majority of the farmers (51.63%) were categorized under 'low' annual income category. The respondents under this category earned in between Rs. 50,000 and Rs. 4,50,000 annually. The farmers who were classified under 'medium' annual income category accounted for 29.67 per cent. They earned between Rs. 4,50,000 and 8,50,000 annually. On the other hand, only 18.70 per cent of the respondents were classified under 'high' annual income category who earned between Rs. 8,50,000 and Rs. 12,50,000 per annum.

> Almost all the respondents (98.78%) possessed larger milch animals including buffaloes and cows, especially buffaloes. On the other hand, poultry animals were kept by 8.1 per cent of the respondents while 4.47 per cent of the total respondents possessed draught animals like oxen. The findings also showed

that smaller milch animals like sheep or goat were reared by just 3.66 percent of the farm respondents.

➤ Maximum number of the respondents (45.93%) practiced specialized farming. On the other hand, mixed and diversified types of farming were found to be practiced by 28.46 per cent and 25.61 per cent of the farmer respondents.

➤ Majority of the respondents (70.73%) had medium information seeking behaviour from personal localite sources for agricultural information, followed by 25.20 per cent of respondents who had high degree of information seeking behaviour from such sources. The remaining 4.07 per cent of respondents showed low level of information seeking behaviour from personal localite sources. The most preferred personal localite sources for agricultural information were relatives, progressive farmers and neighbours.

➤ Majority of respondents (57.73%) showed high level of information seeking behaviour from personal cosmopolite sources regarding agricultural issues. About 35.77 per cent of respondents were categorized under medium information seeking behaviour category. Only 6.50 per cent of respondents showed low information seeking behaviour from personal cosmopolite sources. The most utilized personal cosmopolite sources for agricultural information were input retail shops and company agents.

➤ More than three-fourth of the respondents (75.20%) belonged to medium category followed by 21.95 per cent of respondents who belonged to high category of information seeking behaviour from mass media sources regarding agricultural issues. Just 2.85 per cent of respondents belonged to low category of information seeking behaviour from mass media sources. Television, mobile phones and newspapers were the most utilized mass media sources among the respondents for seeking agricultural information. The farmers also used SMS service and social media platforms like WhatsApp, Facebook, YouTube etc. for sharing agricultural information.

➤ Slightly less than half of respondents i.e. 47.97 per cent belonged to medium category of the agricultural information seeking behaviour from extension methods. They were followed by 39.02 per cent of respondents who belonged to low category. Only 13.01 per cent of respondents were found to be in high category of agricultural information seeking behaviour from extension methods. Kisan Mela and meetings were the most preferred extension methods for seeking agricultural information by the farmers.

➤ In context of overall information seeking behaviour regarding agricultural issues, maximum percentage of the respondents (51.63%) had medium information seeking behaviour followed by 27.64 per cent of respondents who had high information seeking behavior. Only 20.73 per cent of respondents showed low information seeking behavior regarding agricultural information.

➤ Almost half of the farmers (49.19%) exhibited medium level of innovativeness. Meanwhile, 28.05 per cent of the respondents possessed high level of innovativeness and 22.76 per cent of the farmers showed low level of innovativeness.

➤ Maximum number of respondents (47.15%) possessed medium level of risk orientation. On the other hand, 30.49 per cent farmers showed high level while 22.36 per cent farmers showed low levels of risk orientation.

➤ Majority of the farmers in the study area (57.32%) showed medium level of scientific orientation, followed by farmers belonging to 'high' category of scientific orientation who constituted 25.61 per cent of the total sample. It was also found that 17.07 per cent farmers exhibited 'low' level of scientific orientation.

➤ Maximum number of respondents were categorized under high category of ecological consciousness (45.53%). It was also found that more than a quarter of respondents (26.42%) showed medium level while around one-fifth of the respondents (19.92%) exhibited low level of ecological consciousness.

➤ Maximum number of respondents (42.28%) possessed high economic motivation. Further, it was observed that 36.99 per cent of the farmer respondents were categorized under medium and 20.73 per cent respondents were placed under low economic motivation categories.

➤ In context of awareness about alternative stubble management measures, the majority of respondents (97.96%) were aware about Happy Seeder/Zero tillage, followed by thatching (89.43%), in-situ incorporation (86.17%) and use of stubble as mulching material (71.54%). The rest of the measures in the order of decreasing level of awareness were found as: use of stubble in packing material (58.13%), mushroom production (38.21%), fuel purposes (28.04%), biogas production (25.20%) and bio char production (23.17%). These alternatives were followed by other stubble management measures like compost making (13.82%), charcoal production (6.50%) and methane gas production (5.28%).

➢ Maximum number of respondents (45.93%) were having medium level of awareness about alternative stubble management measures. On the other hand, about 29.27 per cent of the respondents belonged to high awareness level. It was also observed that just less than a quarter of respondents (24.80%) were classified under low awareness category.

➢ Majority of the farmers (67.48%) were having cropping intensity between 182.95% to 188.67% and categorized under medium category. The farmers having low cropping intensity, i.e. below 182.95% constituted 19.11 per cent of the total sample. On the other hand, 13.41 per cent of the farmers had high cropping intensity, i.e. more than 188.67%.

5.1.2. Stubble Burning Behaviour of the farmers

➢ Majority of the respondents (51.63%) showed positive attitude towards stubble burning, followed by neutral (35.37%) attitude and negative (13.00%) attitude towards stubble burning.

➢ Maximum number of respondents (42.68%) possessed strong level of subjective norms associated with stubble burning. On the other hand, 41.47 percent of farmer respondents had moderate level whereas 15.85 per cent of them had low level of subjective norms associated with stubble burning.

➢ Nearly sixty per cent of the respondents (58.54%) showed 'high' degree of perceived behavioural control related to stubble burning, followed by farmers showing 'medium' (27.23%) and 'low' (14.23%) degrees of perceived behavioural control related to stubble burning respectively.

➢ Maximum proportion of respondents (41.06%) were categorised under 'medium' level of readiness or intention to burn the stubble after harvesting. It was further noted that almost equal proportion of farmers were categorized under 'high' category (40.24%). Lastly, those farmers who were categorised under 'low' category of behavioural intention regarding stubble burning constituted only 18.70 per cent of the total respondents.

➢ Majority of farmer respondents were engaged in stubble burning every year in the last five years, thus classified under 'very high' frequency (58.13%). Around a quarter of farmer respondents (24.80%) burnt the paddy stubble after harvesting in four out of five years while 9.76 per cent of the farmer respondents had burnt the stubble in last three out of five years, thus categorised under 'high' and 'medium' frequency categories of stubble burning respectively. It was also revealed that 4.06 per cent of the farmer respondents burnt the stubble on field twice in last five years while 3.25 per cent of the respondents had burnt the stubble just once during the time period which were put under 'low' and 'very low' frequency categories of stubble burning respectively.

➢ Almost half of the respondents (49.19%) burnt more than 80 per cent of the total stubble produced on their field and were classified under 'very large' proportion of stubble burnt category. They were followed by the farmers who burnt 60-80 percent of the total stubble produced, thus classified under 'large' proportion category. The farmers who burnt 40-60 per cent (medium), 20-40 per cent (little) and less than 20 per cent (very little) proportions of the total stubble produced on their fields accounted for 28.86 per cent, 10.57 per cent, 6.10 per cent and 5.28 per cent respectively.

➢ In context of overall stubble burning behaviour, more than half of the farmers showed 'high' performance of stubble burning behaviour (53.66%) followed by 'medium' (26.42%) and 'low' (19.92%) performers of stubble burning behaviour respectively.

➢ An interesting observation was made on the pattern of stubble burning by the farmers. Farmers either burnt the stubble partially or completely. In the partial stubble burning, only the stubble left on trail by the mechanical combine harvester along its line of movement is burnt. It was done for the purpose of killing the pests and the remaining stubble is mixed into the soil. On the other hand, the complete burning of the stubble after harvesting kills the pests that are supposed to be thriving on the stubble and saves efforts, fuel and time. Complete burning also facilitates smooth land preparation operations for subsequent wheat crop.

6.1.3. Extent of influence of attitude, subjective norms and perceived behavioural control related to stubble burning on behavioural intention regarding stubble burning

- ➤ The Structural Equation Modeling was used to analyse the hypothesised relationships between the Theory of Planned Behaviour constructs. The measurement and structural models were found to be adequately fit according to the standard fitness indices which were calculated using AMOS software. After that the estimation of structural model was done.

- ➤ The standardized path estimate obtained for the path between attitude towards stubble burning and behavioural intention regarding stubble burning was 0.61. The p-value obtained was 0.012, which is less than 0.05. Thus, it was found to be significant at 5 per cent level of significance. It meant that the behavioural intention to burn the stubble increased 0.61 units for a unit increase in the attitude towards stubble burning. The relationship between attitude towards stubble burning and behavioural intention regarding stubble burning was found to be positive and significant. Thus, the null hypothesis H_{01} was rejected.

- ➤ The estimated value of standardized path coefficient for the path between subjective norms and behavioural intention was found to be 0.28 and the p-value obtained was 0.047, which is less than 0.05. Thus, this relationship was found to be significant at 5 per cent level of significance. It meant that for a unit increase in the subjective norms associated with stubble burning, there was an increase of 0.28 units in the behavioural intention of the farmers regarding stubble burning. The relationship between subjective norms associated with stubble burning and behavioural intention regarding stubble burning was found to be positive and significant. Thus, the null hypothesis H_{02} was rejected.

➤ The standardized path estimate value of the path between perceived behavioural control and behavioural intention regarding stubble burning was 0.47. The data also revealed that p-value obtained for this path was 0.031, which was less than 0.05. Therefore, the relationship was found to be significant at 5 per cent level of significance. It implied that for a unit increase in perceived behavioural control associated with stubble burning, there was an increase of 0.47 units in the behavioural intention regarding stubble burning. The relationship between perceived behavioural control and behavioural intention regarding stubble burning was found to be positive and significant. Thus, the null hypothesis H_{03} was rejected.

6.1.4. Extent of influence of behavioural intention regarding stubble burning on stubble burning behaviour of the farmers

➤ The path between behavioural intention regarding stubble burning and the stubble burning behaviour of the farmers was estimated with the standardized value of 0.53. The obtained p-value for this path was 0.018, which was less than 0.05. Thus, the relationship between the two variables was found to be significantly positive at 5 per cent level of significance. It meant that for each unit increase in behavioural intention regarding stubble burning, there was an increase of 0.53 units of stubble burning behaviour.

➤ The null hypothesis $H_{04,}$ which stated that there was no significant relationship between the behavioural intention and stubble burning behaviour of the farmers, was rejected.

6.1.5. Relationship of profile characteristics with stubble burning behaviour of the farmers

➤ The variables like age, size of land holding, livestock possession, innovativeness, risk orientation, scientific orientation, ecological consciousness, awareness regarding alternative stubble management measures and cropping intensity, annual income and information seeking behaviour had non-significant relationship with the stubble burning behaviour of the farmers.

➢ Economic motivation was the only variable to have a positive and significant relationship with the stubble burning behaviour of the farmers. It showed a relationship of moderate strength with the dependent variable.

➢ The variables with grouped data like gender, education and type of farming showed independent relationship with stubble burning behaviour of the farmers.

6.1.6. Social Marketing plan

➢ The social marketing plan was proposed on the basis of rest of the findings of the study using the BEHAVE framework. The plan was divided into two components; the first addressing to the attitude change and the second considering the overall stubble burning behaviour.

➢ The bottom line of the plan focussing on attitude change was "changing the attitude of the farmers regarding stubble burning by making them realise the opportunity cost of low-cost stubble management alternative."

➢ The farmers belonging to 'very high frequency' category of stubble burning and those with semi-medium land holding category were selected as the target audience.

➢ The strategy focussed on making the farmers aware of the benefits they are missing-out on by burning the stubble and how Pusa Decomposer can help them reap those benefits using different persuasive techniques.

➢ The media mix for conveying the messages included television, mobile phones and newspapers. The formats chosen were text messages, videos, digital images/posters, etc. and the messages would be designed in regional language.

➢ The messages designed for the promotion of the product would be pre-tested among a small representative target audience via pilot testing.

➢ These should be followed by implementation of the planned interventions, evaluation using pre and post-tests by attitude scales prepared in advance and refining of the process wherever required.

- The bottom line of the plan focussing on overall behaviour change was to "reduce the number of stubble burning incidents in comparison to last season in Malwa region of Punjab."

- The target audience for the proposed social marketing plan will constitute those farmers who showed very high frequency of stubble burning and possessed semi-medium land holding, i.e. between 5-10 acres.

- The strategy will focus on the following points: (i) introduce a low-cost alternative stubble management measure, (ii) provide a good compensation/incentive to those who do not burn the stubble, (iii) use emotions to remind the farmers of ill-effects of stubble burning at regular intervals using various methods, and (iv) make the farmers aware of benefits of not burning the stubble.

- The four Ps of social marketing will consist of: (i) Product: Pusa Decomposer, (ii) Price: Rs. 20 per 4 capsules, (iii) Place: Input retail shops/company agents, (iv) Promotion: TV, SMS, social media.

- The pre-testing of the messages regarding the awareness and use of the product could be performed among the target audience for necessary modifications. For pre-testing, the farmers could be gathered in focus-groups to discuss the impact of the messages and product on their perceived barriers and benefits.

- The implementation of the plan will feature the use of lessons learned during prototyping and pre-testing of the product and messages. All the decisions made during previous steps are made sure to fall in respective places. It also features continuous reproduction of the product and the messages that are made to create awareness.

- The final evaluation, i.e. change in the number of stubble burning incidents would be traced by the Consortium for Research on Agro-Ecosystem Monitoring and Modeling from Space (CREAMS) at Indian Agricultural Research Institute (IARI), New Delhi which regularly monitors the active fire events due to stubble burning at regional scales. Apart from these, it is equally important to assess the awareness, attitudes and perceptions of the target audience regarding the proposed product. For these, different scales, indices and knowledge tests can be constructed beforehand. The effectiveness of the dissemination channels should also be a part of evaluation process. The sale of product units from the point of sale could be an indicator of the effectiveness of dissemination channel. The evaluation schedule should be strictly adhered for both on-going and terminal components for fruitful and timely fine-tuning.

> ➢ The last stage of BEHAVE framework is the refining of the campaign, based on the results of the evaluation. For this purpose, a certain time is set aside to re-evaluate whether the plan is working or not. The refining should be based on the answers to the questions like whether the picked target audience is right, whether there is any change in their behaviour, how right are the chosen barriers and benefits. It should also address the fact that whether there are any unintended consequences of the interventions, whether the messages are getting through the target audience, etc. Everything should be checked, and necessary refining should be done.

5.2. CONCLUSION

The study was conducted using the Theory of Planned Behaviour for finding out the determinants of stubble burning behaviour of the farmers in Malwa region of Punjab. Findings of the study and the logical interpretations of their meaning in the light of other relevant facts prompted the researcher to draw the following conclusions:

Most of the farmers preferred burning the paddy stubble after harvesting the crop so that the subsequent wheat crop could be sown on time. Some alternative stubble management measures like use of Happy Seeder, Super Seeder, etc. were available but only a few farmers could afford them. Most of them did not get satisfactory results and thus, reverted back to burning the stubble. From the findings, it can be concluded that majority of farmers in the study area belonged to middle age category i.e. 26 to 52 years, were males, educated up to intermediate level, possessed semi-medium land holding (5-10 acres) and earned between Rs. 50,000 to Rs. 4,50,000 per annum. Most of them had kept large milch animals like cows and buffaloes and were practising specialized farming.

Most of the respondents were classified under medium level of information seeking behaviour. The most reliable sources of agricultural information were found to be relatives, input retail dealers, TV, mobile phones (SMS and social media), Kisan Melas and meetings. Most of the farmers possessed medium level of innovativeness, risk orientation and scientific orientation while high levels of ecological consciousness and economic motivation. It was also revealed that most of the farmer respondents showed medium level of awareness towards various stubble management measures and had medium cropping intensity.

The findings related to constructs derived from Theory of Planned Behaviour were concluded as: the maximum respondents showed positive attitude towards stubble burning, faced strong subjective norms associated with stubble burning, had high perceived behavioural control and medium level of behavioural intention regarding stubble burning. Majority of farmers showed 'very high frequency of engagement' in the stubble burning behaviour, i.e. they have been burning the paddy stubble on their field since last five or more years. It was also revealed that most of the respondents burnt 'very high proportion', i.e. more than 80 per cent of the total stubble produced on the field. The majority of farmers showed 'high' performance of the overall stubble behaviour.

The present study made an attempt to analyse the relationship and extent of influence among Theory of Planned Behaviour constructs. The Structural Equation Modeling was used for this purpose using AMOS software. The results of structural model estimation revealed that out of attitude, subjective norms and perceived behavioural control, the attitude of the farmers towards stubble burning was the most influential factor in the formation of behavioural intention regarding stubble burning. All the hypothesised relationships between the Theory of Planned Behaviour constructs were found to be positive and significant. It was also revealed that behavioural intention regarding stubble burning influenced the stubble burning behaviour of the farmers. The relationship between behavioural intention and stubble burning behaviour of the farmers was also found to be positive and significant.

The results regarding the relationship between profile characteristics and stubble burning behaviour revealed that variables like age, size of land holding, livestock possession, innovativeness, risk orientation, scientific orientation, ecological consciousness, awareness regarding alternative stubble management measures and cropping intensity were positively and non-significantly correlated with the dependent variable. The variables like annual income and information seeking behaviour were found to have negative and non-significant correlation with the stubble burning behaviour of the farmers. The variables with grouped data like gender, education and type of farming showed independent relationship with stubble burning behaviour. The economic motivation was the only variable which showed positive and significant relationship with the stubble burning behaviour of the farmers.

Therefore, based on the findings, it could be concluded that the rising number of stubble burning incidents are not getting under control due to lack of affordable alternative stubble management measures and behavioural determinants also play a key role in influencing the stubble burning behaviour of the farmers.

5.3.   RECOMMENDATIONS AND IMPLICATIONS

Based on the findings and conclusions of the study, the following recommendations and implications were made:

➢ Among the Theory of Planned Behaviour constructs, the attitude of the farmers towards stubble burning was found to be the strongest influencing factor on their behavioural intention. Thus, it is highly likely that a change in attitude and other determinants will result in change in behavioural intention and ultimately in the stubble burning behaviour of the farmers. Therefore, the strategy to change the attitudes of the farmers towards stubble burning might be the first step in changing the overall behaviour of the farmers regarding stubble burning.

➢ The attitude, subjective norms and perceived behavioural control were found to affect the behavioural intention of the farmers. The study only proposed a plan which focuses on changing the attitude of the farmers towards stubble burning. Separate strategies may also be formed for changing subjective norms and perceived behavioural control, which ultimately would bring change in behavioural intention and stubble burning behaviour of the farmers.

➢ Since the Theory of Planned Behaviour was found to successfully explain the stubble burning behaviour of the farmers, it is recommended that attempts can be made to explain other farmers related behaviours in agriculture.

➢ The social marketing plan suggested in the study can be tested further by researchers, policy makers and extension functionaries for its wider application.

➢ It was found that most of the farmers were ready to give up stubble burning in case of availability of a low-cost alternative. In this context, the study also suggests a low-cost alternative, Pusa Decomposer. It can be pushed through various mass media and other agricultural information seeking channels utilized by farmers as suggested in the study.

➢ The findings of the present study can be utilized by government organizations or private agencies to develop action plan top reduce the stubble burning incidents.

5.4. SUGGESTED AREAS FOR FUTURE RESEARCH

A small piece of study as has been conducted cannot provide all the necessary tools and information for stopping or reducing the stubble burning incidents. Therefore, the following suggestions are made for further studies:

➢ The present investigation was confined to understanding the determinants of stubble burning behaviour of the farmers based on Theory of Planned Behaviour only. More such behavioural theories could be tested which could explain this behaviour in a better way with more determinants. This might provide even a better understanding of why farmers burn the stubble.

➢ The present study has been carried out among the farmers of Malwa region of Punjab only. So, similar studies may be conducted in other areas and states as well to verify the findings of the present study and in order to have a better understanding about the stubble burning behaviour of the farmers.

➢ A comparative study can also be done in different states focussing on the stubble burning behaviour of the farmers to find varied determinants according to different agro-climatic and social conditions.

➢ An in-depth study on information and training needs of farmers to use various alternative stubble management measures can be conducted.

➤ The present study is confined to studying stubble burning behaviour of the farmers by applying the Theory of Planned Behaviour. It is suggested to explore the application of Theory of Planned Behaviour in areas like adoption of conservative agricultural practices, etc.

➤ Studies focussed on design, development and implementation of social marketing plans may be conducted based on the findings of behavioural determinants in order to tackle the undesirable behaviours and promoting the desirable behaviours related to farmers and agriculture.

www.ingramcontent.com/pod-product-compliance
Lightning Source LLC
LaVergne TN
LVHW010321200726
843507LV00010B/1321